Future Sounds

Future Sounds

The Temporality of Noise

Stephen Kennedy

BLOOMSBURY ACADEMIC
NEW YORK • LONDON • OXFORD • NEW DELHI • SYDNEY

BLOOMSBURY ACADEMIC
Bloomsbury Publishing Inc
1385 Broadway, New York, NY 10018, USA
50 Bedford Square, London, WC1B 3DP, UK

BLOOMSBURY, BLOOMSBURY ACADEMIC and the Diana logo are trademarks of Bloomsbury Publishing Plc

First published in the United States of America 2018

Cover design: Louise Dugdale
Cover image © Stephen Kennedy

A catalog record for this book is available from the Library of Congress.

ISBN: HB: 978-1-5013-2105-4
ePDF: 978-1-5013-2106-1
eBook: 978-1-5013-2107-8

Typeset by Newgen KnowledgeWorks Pvt. Ltd., Chennai, India

To find out more about our authors and books visit www.bloomsbury.com and sign up for our newsletters.

For Gemma Lori & Erin

Contents

Preface

Future Sounds is a book about time and about history. It also explores the possibility of extricating ourselves from the dialectical bind that keeps the audible and the visual separate but related in ever shifting ratio. Escaping from this bind, it is proposed, will allow for a more nuanced and inclusive engagement with many of the urgent questions that are present in the digital age. By employing the figure of noise, defined as the presence of everything all at once, the strictly linear account of time and historical progression can be avoided. In its place will be a multisensory, multitemporal account where established dualisms are forced into ever-closer association if not collapsed altogether.

Notions of linearity and temporal succession will be juxtaposed with a series of reflections that both challenge and augment Jaques Attali's thesis that music is a herald and that it precedes that which will later become manifest in the realm of political economy. While Attali offers a brilliant historical analysis of music and its significance in terms of critical theory, his is a model that cannot fully account for the complex non-linearity of the contemporary digital age. Attali's model is also unable to reconcile the concerns expressed by a disparate range of theorists who are uncomfortable with the anthropocentric trope that falls back on correlationism as a way of understanding the world as it is, the world as it appears to be and the world as it should be.

Like most theoretical ventures, this book, in its attempt to correct the failings identified in current modes of analysis, is bound to ask more questions than it delivers answers, but it makes no apologies for that. Not least will it question the place of humans in the contemporary universe that vibrates incessantly creating patterns and systems that, while they resolve to calm the ensuing chaos, ultimately serve only to confirm its complexity and unpredictability.

Introduction

This book sets out to explore the proposition that there exists a realm, or a temporal dimension, where infinite formulations occur and reoccur, creating patterns that can be taken to constitute the entirety of our universe. Such patterns are always in motion. They are part of an ongoing process that is chaotic and evasive, and as such they can only be taken as constitutive if fixed notions of certainty are brought into question and if distinctions between the real and the virtual, the subject and the object, and between the world as it is and the world as it appears, are collapsed. Events in this realm and their perception are not separated by critical distance, but instead operate as a singularity. This proposition suggests that patterns occur in such a way that, if understood correctly, can facilitate a new and informed grasp of the unfolding temporal continuum as it is conceived of in contemporary digital environments. Supporting this proposition will require a clear account of both existential and historical modes of time. Distinguishing between these two modes will allow technology's impact on the way time is experienced individually and collectively to be gauged, and will also facilitate an assessment of our (humans' and non-humans') ability to reliably predict what is yet to come.

This unfolding temporal continuum can be best conceived of as a sonic realm, or, as a realm composed of noise. This is because its evasive mobility resists the fixing of the gaze that is a characteristic of visual methods of analysis. The sonic can also serve, according to some, as an early warning system for things as yet unseen. In order to assess any such premonitory characteristics, noise, sound and music will enter into complex negotiations with a range of political, economic and cultural assumptions.

To achieve its aims, the book draws on a range of disciplines to account for the relationship between the digital and the temporal. In doing so, it will measure the impact such an account has on the future prospects for rational thought in a time period when complexity and uncertainty prevail. Philosophy, cultural theory, musicology, literature, politics and economics combine with science to explore the idea of a 'digital paradox' where chaos and entropy as aspects and measures of the digital age are calmed and stabilized by digital processes themselves, in a *post*-political economy where multiple creative interactions in a vastly expanded system trigger new forms of social and political engagement. The temporality of this *post*-political economy, if indeed this is where we are, cannot be understood via recourse to existing methods but must engage notions of genealogy as pattern, and noise reduction as compression, where the removal of unnecessary or inconvenient information is a means of ordering chaos. It will consider this 'reduction of complexity' as a means of system formation where archaeologies and genealogies of sound and noise, rather than dialectics, are the dominant model employed to make sense of the contemporary digital time period.[1] This will in turn lead to questions relating to why/how the digital operates as a sonic spatiotemporal environment. It will also address the limitations of currently available analytical tools to account for it. Conventional discourse analysis, archaeology and genealogy can only account for that which has been inducted into the system, that which is present as historical evidence – the visible. Michel Foucault showed us how to delve into such evidence, and how to take account of complex twists and turns in history, and how to search for openings that would facilitate events that did not fit the formal narrative of visual and dialectical logic. Now there is a need to go further, because twists and turns may be all there is. If so, then the viability of an archaeology of noise and sound will need to be explored. For Foucault archaeology explained the patterns that had been formed as discourse, but not the movement of the changing patterns. For this he resorted to genealogy and a rhythm of discontinuity in part inspired by Henri Lefebvre. This will be developed and extended here, asking: Why not a singular continuity where rhythm is not marked by metric breaks but by syncopation and ultimately noise (as everything at once)?

For Jaques Attali (1977), noise (later organized as music) operated as a resistant force that came before more general political and economic change. As such it could potentially be used to predict future events. The plan here is to invoke Attali's thesis in relation to contemporary music and also to expand on his use of the term noise so that it can be used to account for a less predictable and constantly dynamic cosmos like the one described by Alfred North Whitehead, and more latterly by Steven Shaviro and Greg Hainge among others. Such an expanded definition of noise embraces all the other senses but is prefaced here on a sonic sensibility. If for Jean Luc Nancy (2007), the visual had infected listening, then here the sonic infects the entire sensory realm and beyond. For this is not a phenomenological account of noise. Noise is a fluid set of always forming, re-forming in-forming practices and events that are complex, fuzzy, connected and multitemporal, moving seamlessly between and among subjects and objects.

Where most scholars to date have set out to place sound on an even footing with the visual, the intention here is to explore the possibility that the sonic, and noise more specifically, might support not a claim to equivalence, but a radical departure for critical thinking more generally. This comes at a time when critical thinking is trying to negotiate the complexities of an algorithmic digital economy that produces information and knowledge as an unrelenting torrent, moving through and between space(s) where notions of the real and the virtual are no longer easily defined. The practice of accounting for such torrential complexity has brought with it questions pertaining to the privileged status of human beings within such contested space(s). Many of these questions have arisen out of a 'speculative turn' that postulates the existence of distinct spatiotemporal realms: the realm of subjective experience and the realm of objective reality where phenomena are not dependent on human presence or perception for their ontological validation (Shaviro 2014). Noise, unlike sound however, in so far as it is not limited to human perception, offers a unique means by which we can surmount this bifurcation and explore the competing hypothesis of a singular vibrating 'one substance cosmology' (Kennedy 2015) that can support notions of the real without recourse to a separate speculative realm. Employing the figure of noise in this way will allow a significant contribution to be made to these important and timely debates.

The interplay of time and space will be a constant concern in what follows. In his book *The Information Bomb (1998)* Paul Virilio announced – as a counterpoint to Francis Fukayama's famous declaration of 'the end of history'

(1989) – 'the end of space', or 'the end of geography', and the continued importance of time in relation to digital technology. This volume interrogates this idea and critically examines the relationship between the digital and the temporal. In so doing it will consider whether a detailed analysis of noise as a temporal phenomenon can reveal properties that make it uniquely appropriate to this task. Noise as a category will be engaged with in such a way as to invoke the always related and sometimes contrasting categories of sound, vibration and music.

The claim relating to the 'end of geography' (space) is accompanied in *The Information Bomb* by an increased *focus* on speed, and with that our attention turns neatly to time and to history. But must that attention be bound by the act of *focusing* (Virilio 1998: 9)? Ideas relating to the shift from the visual to the audible (the anti-ocular turn[2]) have been in circulation for some time – in my own work and in that of many others – but so far, I would claim, without any real impact. The plan here is to move further towards what has been described elsewhere as a 'sonic-economy' (Kennedy 2015), which can facilitate new kinds of thinking and with it new kinds of knowledge pertaining to our digital existence. In what follows, it will be necessary to consider the relationship between time and technology and ask whether the temporal nature of noise and sound (if indeed they are temporal in nature) can be usefully harnessed in developing a critical understanding of our technologically mediated past, present and future.

This question relating to the temporality of sound will be addressed in relation to Jonathan Cohen's claim that sound when understood in relation to other sensory phenomena is not uniquely temporal. Ultimately this discussion informs the argument relating to the temporal status of *noise* and will operate as a factor in determining whether it possesses premonitory characteristics that can be usefully employed in plotting a future trajectory for social, political, economic and technological formations. This in turn will require that different notions or modes of temporality be considered within the digital context, as the extent to which a significant shift has occurred is brought into critical range (Barker 2012). The different modes are: continuous and discontinuous, time in relation to analogue and digital processes, time as speed related, existential time as it is experienced, and historical time as the transition between defined temporal periods. In considering them, care will be taken to pay attention to the implications of this for the efficacy of dialectical thought more generally and particularly in its ability to account for the digital.

Another key point of consideration will be the temporal nature of the digital itself as an always shifting reconfiguring onflow, to borrow Nigel Thrift's phrase (Thrift 2008), that enacts *representation-as-compression*, where information is *processed* as the digital and the analogue move into and out of each other seemingly without impediment. Indeed, representation itself might be regarded as a process of compression, where meaning is both deduced and reduced through reason to create a formal grammar of sensible reality that extracts and abstracts the 'unreasonable' and discards it. This happens in such a way that can itself never be truly represented as static certainty because the information that is not included in any compressed representation does not cease to exist.[3] In fact its enduring presence, albeit too often ignored, serves to highlight the process of representation and its fragility when diverted through the figure of noise. In this sense, the MP3 serves as a useful analogy for the idea of *representation-as-compression* as a feature of the digital age, and this will be demonstrated later through a discussion of Jonathan Sterne's work in this area.

The digital might further be understood as a non-linear information flow that does not need to be confined to real or virtual categories, because as a temporal rather than spatial entity it escapes the need to be defined ontologically.[4] The movement of elements from analogue to digital and back again will be explored as characteristics of a system that demonstrates both wave and particle features simultaneously, as a *datawave* that traverses complex media environments. These elementary constituents are similar to Pierre Schaefer's sound objects.[5] The *digital* as an amalgamation of non-extended 'sound objects' reconfigures itself constantly in a universe where things are always new and never temporally stable. To understand this, it will be helpful, at particularly apposite moments, to go back to Alfred North Whitehead and to the question of creativity as a process of constant renewal and innovation. This will be contrasted in due course with more contemporary theoretical ideas, predominantly those of Graham Harman, in relation to his conceptualization of newness and creativity.

Noise is deployed here as a means of engaging with the complexity of the digital realm as it specifically opens up new ways of approaching political and economic events that are relative to but not necessarily caused by the conditions of their existence. Further, the analogical deployment of noise to help demonstrate this shows how creative engagement need not always *oppose* dominant forces but how it can pursue alternative strategies and opportunities, supported by the idea of difference as an expressive force, to open up multiple opportunities and options. Political economy, as it is conceived of here, operates within this analogical sonic environment in so far as it is regarded as relative to all other aspects within a dynamic universal realm and not as a specific set of compressed and formally categorized activities.

This volume strives not towards a diagnosis but a prognosis. The intention is to explore the possibility of developing a means of classifying and tracking noise as it emerges as music within a complex system creating scenarios for back testing in relation to retrospective historical case studies. It will consider whether computer modelling can quantitatively link sonic phenomena so as to finally demonstrate how musical change foreshadows broader social, economic and technological change as Attali claims it does?

One very important discussion, necessary in terms of addressing the issue of predictability, will revolve around dialectics and the ability of that system of analysis to account for contemporary conditions and future developments. This discussion further relates to the question of whether noise, sound and music can be usefully thought of dialectically. This question has a long history and it will be examined in detail in Chapter 2 with specific emphasis on a critique of the idea that noise is in essence a violent form of resistance that opposes prevailing logic.[6] If dialectics is always based on negation or even the negation of the negation, then the idea of oppositional practices as central, both politically and economically, and also aesthetically, will need to be carefully considered in contradistinction to noise as a figure through which these practices are rearticulated as non-linear and non-teleological.

Distinguishing between noise and sound raises a number of specific and important issues. Sound as it emerges out of noise as language either remains tethered to the idea of the natural and to presence or alternatively becomes dialectically opposed to noise, and in terms of extracting it from the natural, takes its place within the phenomenology of Heidegger and Derrida as it moves towards absence through technology. This distinction between presence and absence is significant. Sound was initially privileged in metaphysical terms as it presupposed presence, only to be displaced by the visual as critical distance through representation came to dominate. This is important in terms of temporality and the established idea that plots the movement from orality through the literate to electronic consciousness (Sterne 2011). This trajectory is itself significant if electronic consciousness is taken to apply to the digital age, for it describes a historical progression from a privileging of sound and the oral to the primacy of the visual and now back to sound as the emphasis on presence and absence is mediated in a circulatory manner. Such circulatory mediation links in turn to some fundamental phenomenological ideas: sound, orality and presence (as human and cultural as distinct from natural), incline towards a kind of sensual engagement that abstracts Being and ultimately renders it unknowable. Absence on the other hand facilitates a degree of separation and reflection that can be deployed as a methodological form of engagement unhindered by direct sensory experience that can reveal truth as Being. Being in this sense is a term that denotes the essential nature of existence as a particularly human mode of questioning. For Heidegger, it had become a theoretical concept because it could not, according to Kantian logic, be encountered without sensory mediation. Being itself was not accessible and therefore could only be speculated on as noise, becoming sound only when it *made sense*. This abstraction of Being was not acceptable to Heidegger, or Derrida, for whom a particular mode of questioning not bound by presence could reveal Being. Their approach was prefaced on the instability of the written word as key to the formation of a mode of enquiry that did not rely on the division of reality into the world as it is and the world as it is to us. In a move towards absence, the deconstruction of language opened up for them a clearing where Being revealed itself as essential truth. Paradoxically then, noise, as that which has been abstracted, returns as the hitherto inaccessible or incomprehensible is revealed, but this time in its silent contemplative linguistic form accessible only to human subjects.

The limiting nature of the subject–object dualism and the theoretical abstraction of Being that Heidegger noted, are central tenets of dialectical thinking, and the intention is to challenge them in what follows. They support the imposition of a rational logic that is able to compress and represent the world as understandable only within a specific framework. Everything outside of that is somehow deemed to be an irrational anomaly – noise to be quietened and soothed. To a degree Heidegger and those who followed him did away with such models but were not able to fully dispense with the idea of a fundamental ontology that was itself timeless, though historically contingent in so far as prevailing technological conditions mitigated against accessing essential truth. To counter this, Heidegger sought a reinterpretation of Being, as a reconstitution of a pure form of existing that would serve humankind for eternity. But if the return of noise occurred as a result of Heidegger's challenge to metaphysical abstraction, it did so at a cost. To achieve it, he was prepared to countenance an enforced volkisch movement – a kind of nostalgic return that could itself be interpreted as dialectical. His thought was the antithesis of enframed instrumental thinking that he identified as being so problematic and as central to technological rationality. If his critique of technology was intended to arrest or alter its line of travel however, he was largely unsuccessful and we now find ourselves once more having to find modes of engagement that are appropriate to our contemporary condition.

For Heidegger Being was invisible, and this is significant here. It revealed itself as a clearing only under certain very particular conditions of questioning that could be interpreted as inclining towards a sonic methodology. It would not reveal itself as a visual phenomenon that could itself be perceived (not as a circular phenomenon in the way that Merleau-Ponty described). Rather 'Just as a guitar string will "sympathetically resonate" with a string of the same pitch that is vibrating on a guitar nearby, perhaps the process of Heidegger's questioning allows the Being of the questioner to sympathetically resonate with the awesome and amazing phenomenon of Universal Being'. (Watts 2001: 23). So, Heidegger is useful in providing the critical underpinning that can support a level of engagement with the world beyond subjective perception of objects. It hints at a degree of immersion that is acoustic rather than visual, as the previous quote shows. But that is all it is – a hint. There is no direct acknowledgement of noise, sound or listening as being able to facilitate a clearing that does not require recourse to notions of the inside or outside. In order to directly identify such an acknowledgement, we must move to the ideas that have been developed by Jean Luc Nancy, and his work will be addressed in more detail later. For Nancy, phenomenological engagement with the world is better understood via recourse to listening where an engagement means always-never-being-able-to pin it down. This is specifically described as a 'fading into permanence', where the noise of what cannot quite be grasped or understood is always significant.

Deploying the figure of noise through a reading of Heidegger does allow us to move beyond the visible and the present-to-hand conventional subject/object duality, but in asserting a universal Being accessed by his way of questioning, he imposes an ultimate truth – a border, or limit. To go further beyond the visible and rational limit, it is important to understand how Alfred North Whitehead's position contrasted with Heidegger's. Whitehead was interested less in the idea of absolute truth with its limits and thresholds, and more in the unrelenting nature of the new – the always unfolding reconfiguring patterns of nature. This, it will be argued, can be used as a means of liberating noise from its negative confinement, allowing it free range to account for the unaccountable.

With the unaccountable always in mind, noise can be understood as an inclusive and comprehensive means of grasping the complex nature of the contemporary time period. It is a time that does not seem to play by the established rules, where the speed of change sometimes out runs our ability to account for it. *Making sense* of this relationship between time, technology and the new within a temporality of noise requires an almost obligatory reference to Walter Benjamin's seminal *The Work of Art in the Age of Mechanical Reproduction* (2008). From my perspective, the most enduringly engaging aspect of Benjamin's essay is the extent to which technological means of creative production precede our ability to critically evaluate them. Such a critical lag means that we may always be destined to be asking inappropriate questions, or the right questions at the wrong time. How is it possible, it should be asked, to move forward when there is so much unfinished business? With the modern project still incomplete, it seems we still have not caught up with the postmodern mutation, let alone the post-postmodern or the alter-modern (Bourriaud 2009). What does this tell us about time and our status in relation to it? Are we lagging behind or racing ahead – never in the present? What if we could change this? What if we no longer desire *to make sense?* What might our critical tools be composed of in such a complex temporal scenario? To make such a change requires a radical interpretation of time as a concept in the digital age – one that embraces the 'technological paradox', the post human and noise.

Like the central character Shevek in Ursula Le Guins's novel *The Dispossessed* (1974), the aim of what follows is to reconcile the nature of time and technology with our ability evaluate it and participate in it in a meaningful and creative way through noise sound and music. 'Shevek, [however], was not very sensitive to drama. He liked the verbal splendour, but the whole idea of acting was uncongenial to him. It was not until this second year in Abbenay that he discovered, at last, his art: the art that is made out of time'. (p. 131)

For Shevek, the answer to his questions in relation to instantaneous communication – the *ansible* – lay in the art that was made out of time, in music, as a dynamic resonating model of the universe. How can a similar analysis here help us to install reinvigorated forms of political thought that are in time with the conditions of their existence? The following initiates a small step towards comprehending this question and to a better understanding of why such questioning is so pertinent in the digital age where reliable measures of time and space are less certain and therefore more difficult to represent. To fully achieve this aim requires a thorough philosophical explication of the concept of time and the related cogency of the temporal arts.

Since Jaques Attali's *Noise: The Political Economy of Music* (1977), there has been a consistent movement towards reclaiming the art that is made out of time as a politically significant realm. Sound in general and noise in particular have become common themes in trying to underpin a new kind of political thinking that does not rely on visual representational logic for its legitimacy. The adoption of sound as a model for thinking something new invariably brings us into contact with the temporal as both a general and specific site of enquiry. In so doing, it also brings us into contact with Henri Bergson and Gaston Bachelard. There will be a more detailed exploration of time in relation to the work of Bergson and Bachelard later, but by means of an introduction: if Bachelard speaks to the discontinuous and discrete bits of the digital world, and Bergson to the analogue with its seamless unfolding, then a decision about which to follow will significantly affect the conclusions arrived at when addressing questions relating to temporality in its digital mode.

Bergson took art as an example to show that duration works in a similar way to music. In *An essay on the Immediate Data of Consciousness*, he explained duration by referring to art, and music in particular saying:

> I shall perceive (the ringing sound and the vanishing sound) one in the other, each permeating the other and organizing themselves like the note of a tune, so as to form what we shall call a continuous or qualitative multiplicity with no resemblance to number. I shall thus get the image of pure duration. (p. 105)

Bergson used the motion of a pendulum as an example to compare duration with melody. When listening to the consecutive sounds made by the pendulum, it is not possible to identify which of the sounds has the effect of inducing sleepiness. When each sound is perceived in the other the series of sounds becomes somniferous. Melody, Bergson says follows the same principle. It is not single sounds in music that demands our attention. 'When several sounds make a unity, in other words a melody, they appeal to us. Bergson argued that this is what duration is about. To sum up, melody teaches us what duration is like' (Keiko 2009).

Gaston Bachelard adopted an altogether different approach. For him the notion of the *new* as a radical break rather than a melodic continuity – also a recurring theme here – was the focus. He identified in scientific discourse a dialectical feature in terms of its openness to challenges that could account for significant paradigm shifts (Grant 2005). Time for him was a series of challenges and ruptures. These challenges were creative moments that enacted discontinuities. They were discrete breaks that could be marked and measured. What is significant here is the extent to which a similar argument might be made to support the turn towards sonic thinking, as a mode of thought equally open to challenge and recognized as being in need of constant readjustment. The efficacy of such an argument will be kept in mind throughout.

But what happens if continuity and discontinuity are not dealt with separately but as interactive components of an integrated approach based on noise, sound and constant sonic renewal? Paradoxically, such an approach might demonstrate how discontinuous and discrete moments or turning points are identifiable within a continuous play of interrelated phenomena that resonate sonically. In Bergsonian terms, immersion in time was creative and political. It was something you did (*Elan Vital*), rather than something that was done to you, and as an existential event it was characterized by continuous and resonating duration, experienced as an unfolding that was not conducive to mathematical segmentation. Yet, like Bachalard's echoing moments that rise and fall without conventional cause and effect, sonic events seem to punctuate time in a decisive manner where all that was once certain is suddenly thrown into question as new norms begin to form. The existential, historical and political aspect of sonic enquiry must operate therefore, in time with both the continuity of Bergson and the discontinuity of Bachelard. Noise, sound and music seem to be equally at home then in both continuous and discontinuous modes of analysis.

If melody 'teaches us what duration is like', then noise teaches us what the continuous discontinuity of digital time is like. It names a process, a field of activity, a ground, an economy wherein the rules are always likely to be broken, where surprise and the unexpected prevail despite the imposition of steadfast and reliable pillars of certainty. Noise is an uncompressed array of potential and opportunity – a positive mode of Being. This proposition will need to be supported and the necessary theoretical underpinning will be provided in the chapters that follow.

Noise to date, as both phenomenon and analogical device has mostly, but not solely, been regarded as auditory, disagreeable and uncomfortable. From this perspective, it can be discussed in terms of power and domination as efforts are deployed to either utilise it or bring it under control. From a different perspective though comes George Prochnick (2011). He is in agreement that noise is disagreeable, but he approaches it from the point of view of resistance and not domination. It is he believes the refusal of the masses to be silenced. Hence, for him noise is regarded as oppositional in character – it is a by-product of another action, in this case protest. Similar to its treatment in information theory, noise attaches itself to a signal and distorts its intended meaning. In this case, however noise does not necessarily enter the signal from outside, but gathers force and amplitude from within the communication channel itself. So, for Prochnik noise operates in a closed environment (the aim here is to open this environment so that it can connect with all of the other environments in a manner more akin to Whitehead than Harman, as discussed later on). For him everything vibrates at its own unique frequency when acted on by an energy not from outside itself but from inside the system. But ultimately, he says, everything desires to be still. Noise for Prochnik is the result of a disturbance that will eventually return to rest. It is opposition that causes the disturbance and with it the noise. From this perspective, noise is understood as having an explicit political function aimed at opposition and disruption. Such a closed dialectical approach to noise however is regarded here as too simplistic. A more thorough and philosophically grounded approach is developed by Greg Hainge who approaches noise, unlike Prochnik, as a process and not something that occurs as a result or as an effect. In terms of its being oppositional in a political sense Hainge argues, one must rethink noise against the dialectical grain and rearticulate it in relation to the operation of an electrical field and the resistance therein. In relation to this, he suggests that the medium of transmission both facilitates and impedes the expression of information at the same time (Hainge 2013: 17). Taking our lead from this position, noise describes an environment analogous to an electromagnetic/political field where resistance and opposition are never dialectically resolved, but are perpetually played out in such a way that gives rise to temporary truces and endlessly reconfigured agendas, where positivity and negativity become the medium through which each travels. Such an environment cannot be understood in terms of rationally configured counterpoints and instead must take account of positive and creative expressions that can be resistant without being oppositional.

Within any given medium or system, or in the interconnectedness of systems, noise describes the dynamic operation of that system, as alternating ratio of interaction and a positive material energy field that forms a virtual ground of potential and possibility. If it is not to be the product of a *resistance against*, in the negative sense of trying to hold back, the tide of one kind or another, but an instance of difference that repeats, a number of things need to be explained. Resistance, as Hainge interprets it, draws us into an engagement with Gilles

Deleuze and his work in relation to difference and repetition. Difference for Deleuze is not a measure of the difference between things as defined by their identity, things that come into dialectical conflict (to produce noise) but like the resistance that Hainge describes, is rather a positive expression of the thing itself as it comes into contact with all other things, and where this incessant dynamic interaction is noise. And this dynamic interaction is alive with repetition. For Deleuze, *repetition* was distinct from the idea of the general that was itself based on notions of resemblance and equivalence. If something can be repeated, he says, it has no equivalent; it is a singularity and cannot therefore be exchanged or indeed represented. To demonstrate this, Deleuze refers to festivals that repeat: He describes this as 'the apparent paradox of festivals': how can you repeat an 'unrepeatable', a thing that is unique? (Deleuze 2008: 2). They pop up at certain moments in time, sometimes regular, on the same date each year but undamaged by the passage of time – each as new as the last, and sometimes at less regular intervals. Each time it is the real thing, and each time it is configured in relation to a wholly different set of circumstances particular to that moment. And it is this kind of repetition, as incessantly new and different, that is under consideration here. It is a repetition that occurs in music in the form of the fugue and it is a repetition that contemporary science shows us is both spatially and temporally counter intuitive – capable of being in two (or more) places at once, and instantaneous. So, repetition is patterned at regular and irregular intervals emerging out of the virtual field of noise as both resonance and dissonance and this becomes more evident in a digital time period where movement and incessant interconnection are constant features. This state of being also creates a very different kind of economy where exchange based on generality, representation and equivalence is no longer sufficient. Such economies in so far as they order and compress value and exchange into a very specific set of regulated practices, leave a vast array of waste or unwanted and discarded material that has traditionally been thought of as superfluous and dismissed as noise. This effectively splits the universe in two, as described above – into the world as it appears and the world as it is – a world that can be ordered, represented and understood, and an abstract world that can only be speculated upon, and never known. For Deleuze this kind compression of sense making with its 'exclusion of the eccentric and the divergent' was a feature of almost all philosophy (1990). Such philosophy however cannot account for the complexities of the contemporary age. Currently the compressed and the uncompressed, figuratively and literally, coexist and both are accommodated in a frantic and noisy exchange – not of representation and equivalence but repetition.

What is important going forward is the idea that sound and music, as specific and acoustic modes of noise, demonstrate resistance strategies not in terms of dialectical opposition but as positive and legitimate acts that at their best serve to unsettle the territorial familiarity and the dominant political order. They are constantly ushering in the new. Noise from this perspective can be used as an analogical device for examining a new ground for political praxis that is not constrained by teleological frameworks. It is a model of thinking that draws on sonic economies of distribution and exchange (Kennedy 2010), ones that question strict notions of equivalence and representation in favour of dynamic resonances in an ever-changing cosmos wherein humans and the things humans do and create are regarded as events in the Whiteheadian sense (Shaviro 2012). The eradication of 'difference between' as initiated by Deleuze then, assists us in addressing the problematic nature of resemblance and equivalence and calls for new economic models that can account for the uncertainty of what something is when it can take multiple forms and adopt superpositions in digital time and space.

Resistance to economic decline and the ideological shifts that occurred during the period under consideration in this book (1977–2017) were repeated amidst the seemingly parallel shift from analogue to digital that was as much a feature of the sounds being produced as it was the industrial transformation that was taking place. Was this shift part of a Bergsonian continuity or a Bachelardian rupture? In many senses, it was both a significant turning point – a departure towards a mathematical (digital) realm driven by creative thinking that might be seen as having initiated something entirely new, and an audible continuity in which a resistance that can repeat did so seamlessly across space and time. This period saw significant resistance to a shifting ideological terrain as Thatcherism tightened its grip on the United Kingdom. It also witnessed a change in musical style as traditional forms of Rock, including the Punk phenomenon that has so often been cited as having signalled a significant paradigm shift (McLary *in* Attali 1977), were supplanted by a kind of technologically inflected electronica that seemed to echo the more general transformation at the political and economic levels (Rutsky 1999). This volume extends this analysis up to the present day. This extended time period will be examined and augmented in more detail here with events being treated not as specifically spatialized phenomena but as temporal ones that may provide clues to future development.

Whilst analysing these events, it will be important not to make claims in relation to incidences of cause and effect, beginnings or ends. Instead it is essential to search for confluences, moments of interaction and collision. Rather than simply *looking* for connections, we will listen for them. Investigating events within an acoustic environment will allow us to extricate ourselves from closed systems of thought. Methodologically then, we will be able to make precise statements without recourse to universal laws.

What has happened in the years since 1977 constitutes an ongoing process of resistance that has been repeated. And this repetition constitutes a particular kind of temporality where the past and the future reside in the present. To understand it examples will be discussed as instances of expressive discursive events that initiated positive political action. Though necessarily limited and inflected with the aesthetic judgements of the author they are drawn from a wide array of styles and locations so as to avoid the categorization of noise as a specific genre, or as a particular oppositional style. It will therefore be necessary to embrace resistance even, or especially, where its status as such is not recognized or is misrecognized through representation. With regards to this it is worth referring briefly to the discussion developed by Marie Thompson and Ian Biddle in their introduction to *Sound Music Affect* (2013) where they point to the difficulties facing commentators and theorists in terms of the role of pop music in the socio-political unrest in London in 2010. They cite the playing of Rhianna and Nicky Minaj at protests as counter intuitive in terms of being recognizable as the music of resistance. This is because it is the resistance as difference that repeats and not the music as representation *of* resistance. An approach that embraces noise in the way it is being deployed here cannot therefore pick and choose its protagonists without falling back into a dialectical trap. But what if anything might usefully be gleaned about this temporality from listening to Rhianna or Nicky Minaj, or indeed any other noise, sound or vibration that sonifies a particular moment in, or movement through, time? The following propose new ways of answering such questions. Opening previously closed systems of thought it will go back to go forward. It will match what happened then with what we know now, and ask whether the template that forms out of such questioning can be applied to future sounds as part of a temporality of noise?

The full proposition that it is intended to develop will be set out as follows:

Chapter One: Critical Temporalities

This chapter begins with an exploration of the notions of speed and time as they are developed in the work of Paul Virilio in his book *The Information Bomb*. It will then move on to examine the stereoscopic worldview of the American analytical philosopher Wilfrid Sellars in order to gauge whether his merging of the concepts he calls manifest and scientific image, can assist us in managing the contemporary dualism of the real and the virtual. Finally there will be a broader critique of dialectics in order to question its efficacy as a critical methodology in the twenty first century.

Chapter Two: Noise and Political Economy

This chapter examines in detail Jaques Attali's seminal text *Noise: The Political Economy of Music*. Forty years after its initial publication, the question will be asked as to whether it still offers us a reliable means of understanding the operation of political economy and its perceived trajectory?

Chapter Three: Remembering the Future: 1977–2017

Picking up where Attali leaves off, this chapter examines whether music itself, and not just the political economy *of* music has been a reliable indicator of future events. It will ask, does music herald? A rapid journey through four decades of music will pause occasionally to examine specific examples that are deemed to be pertinent in relation to the overall aims of this book.

Chapter Four: Continuous Discontinuity: A Non-Linear History of Noise

This final chapter moves from the artist Paul Klee through to computer music. In doing so it proposes that the spirituality of music, captured and ordered as it was by Klee, for use in the visual arts, is akin to the digitization of the analogue as a recurring motif of the contemporary age. Both constitute a methodology of ordering, as a means of attaining and organizing knowledge, and in doing so, this concluding chapter argues, they form a process that is both continuous and discontinuous.

Notes

1 See Niklas Luhmann, 'The Future Cannot Begin: Temporal Structures in Modern Society', *Social Research*, vol. 43, no. 1 (Spring 1976): 130–152.

2 See Adrienne Janus, 'Listening: Jean-Luc Nancy and the "Anti-Ocular" Turn', *Continental Philosophy and Critical Theory Comparative Literature*, vol. 63, no. 2 (2011): 182–202.

3 Wolfgang Ernst makes a similar argument in his *Sonic Time Machines: Explicit Sound, Sirenic Voices and Implicit Sonicity* (2016).

4 In relation to this and for a detailed discussion, see Luciano Floridi's work on digital ontology http://www.philosophyofinformation.net/publications/pdf/ado.pdf.)

5 See https://monoskop.org/images/b/bd/Kane_Brian_2014_Pierre_Schaeffer_the_Sound_Object_and_the_Acousmatic_Reduction.pdf.

6 See http://people.uwec.edu/walkerjs/picturesofmusic/Dialectical%20Development%20of%20Music.html.

1

Critical Temporalities

In *The Information Bomb* (2005), Paul Virilio set out a compelling argument in relation to the task at hand here. He posed a counterpoint to Francis Fukuyama's *end of history* proclamation wherein the dialectical grand narratives of modernism were regarded as having played themselves out having achieved all that they were going to achieve. He raised instead the possibility of *the end of space*, or of the *end of geography*, and of a small planet 'held in suspension in the electronic ether of our modern means of telecommunication' (7).[1] Virilio's claim leads him to consider time rather than space as a means of understanding the rapid expansion and significance of the information society. He asked,

> How are we to conceive the change wrought by computerization if we remain tied to an ideological approach, when the urgent need is in fact for a new geostrategic approach to discover the scale of the phenomenon that is upon us? And we need to do this *to come back to Earth* – not in the sense of the old earth which sustains and nourishes us, but of the unique celestial body we occupy. To return to the world, *to its dimensions*, and to the coming loss of those dimensions in the acceleration not now of history (which, with the loss of time, has just lost its concrete foundations), but of reality itself, with the new-found importance of this world time, a time whose instantaneity definitively cancels the reality of distances – the reality of those geographical intervals which only yesterday still organized the politics of nations and their alliances, the importance of which has been shown by the cold war in the age of (East/West) bloc politics. (8)

In making this manoeuvre, he was advocating a new kind of history, based now on the phenomenon described as 'global time'. This was a kind of time that discontinued previous models of history with their concrete foundations, and ushered in a new kind of aspatial history that had been compressed by temporal instantaneity. Local time had been replaced by global time, and spatially reliable continents had lost their geographical foundations and were now constituted as tele-continents where distance was reduced almost to an irrelevance, and where simultaneity and instantaneity reigned. So, for Virilio, as it was for Ursula Le Guin's Shevek as mentioned earlier, time, or more accurately the speed of transport and communications, had changed the nature of space. It had turned it inside out, making what were once localized internal concerns external and peripheral. And conversely, what were once diverse and diffused global concerns now constituted a concentrated singularity where nothing is or can be separated from anything else.

This rapid shift towards global interconnectivity meant that, in geographical terms, *the real city*, which for Virilio was situated in a very precise place that in turn underpinned the politics of nations, was being supplanted by *the virtual city*. This was a 'de-territorialized meta-city' that would form the basis of a new *metropolitics*, a politics that he thought would be identifiable by its totalitarian or rather 'globalitarian' character. This movement from the real to the virtual was an opening up of previously demarcated space. It was also an opening up of systems of thought that had previously created an inside and outside, order out of chaos and quiet out of noise. The aim here is to follow Virilios's lead and let the redefined noise back in, and in doing so to further develop the idea of a kind of quantum entanglement that can be deployed to make meaningful connections among distinct geographical locations. Such connections should be capable of going beyond the representational logic of visual space, to embrace the fluid and temporal nature of acoustic space as a *sonic economy* wherein uniquely different and singular aspects are repeated in a universe that is configured and reconfigured constantly as Deleuze described (2008).

If finite demarcated spatial certainties, which were once the foundation of modern nation states and dominant models of economic exchange, that is to say modern political economy as a whole, are under threat, what is replacing them, or what has replaced them? Is the whole edifice of what we call 'political economy' so contingent on such spatial geometries and geographies as to render the concept/phrase/description meaningless in contemporary terms? Is time replacing or reconstituting space to the extent to which we need to develop entirely new modes of thinking? Virilio certainly thought so. For him temporality altered spatiality and the visual frames of reference that were used to make sense of it. But temporality itself is of only secondary concern within the framework of his thinking. He is concerned primarily with reconstituting space in the aftermath of the temporal assault. He is correct in citing the temporal as having compressed the spatial to such a significant degree as to warrant wholly new kinds of thinking, but instead of following the temporal line of argument to evoke acoustic or sonic modes of thinking, he attempts to reconfigure the visual in such a fashion so as to prolong its efficacy.

He says, 'The more time intervals are abolished, the more the image of space dilates' (2005: 12). As part of this process everything is illuminated and brought into view as reality, and the more it becomes subject to what he calls planetary grand scale optics. Day and night are no longer distinguishable in what is now the always visible 'world time.' Blinded by the light, he says, other senses must be deployed in a cosmos where optics and sonics merge. This optical–sonic motif is significant and should be approached as a proliferation of discrete and discontinuous digital elements travelling at great speed, flattening and in some cases eradicating, specific spatiotemporal dimensions to create a continuity. As such this global singularity should be acknowledged as being possessed of both continuous and discontinuous features that are present as both wave and particle, as will be discussed in more detail later.

But Virilio seems to do something quite different. Although recognizing a closing in of the duality, he imbues the visual with acoustic or sonic characteristic, rather than allowing both to exist simultaneously.

> This is an *active* (wave) *optics*, replacing in a thoroughgoing way the *passive* (geometric) optics of the era of Galileo's spy-glass. And doing so as though the loss of the horizon-line of geographical perspective imperatively necessitated the establishment of a substitute horizon: the artificial horizon of a screen or a monitor, capable of permanently displaying the new preponderance of the media perspective over the immediate perspective of space. (14)

Having flattened time and space into a continuous universe, false discontinuities are introduced as markers of perspective and aids to orientation. The rapid proliferation of this artificial vision, this virtualization as Virilio calls it, has been brought about by the contraction of geographical distance that in turn was 'brought about by the temporal compression of instantaneous telecommunications'. The result was that the world had been split into two – between the real and the virtual – and to understand it Virilio moves towards a stereoscopic–phonic approach. 'As with stereoscopy and stereophony, which distinguishes left from right, bass from treble, to make it easier to perceive audiovisual relief, it is essential today to effect a split in primary reality by developing a *stereo-reality*, made up on the one hand of *the actual reality* of immediate appearances and, on the other, of a *virtual reality* of media trans-appearances'. (15)

Here Virilio moves tantalizingly close to a sonic approach before once again returning to visual and appearances. His stereoscopic proposal, although it serves to reinforce the real-virtual dualism, does so in a way that at least raises the possibility of simultaneous reception and a merging of the real and the virtual to create a singular phenomenon, even if is not sufficiently developed. As a concept stereoscopy can be explored and further extrapolated with reference to Wilfrid Sellars whose ideas will be set out in detail shortly.

The proliferation of these new kinds of optics served to bring light everywhere, eradicate the shadows, and re-energize that which had previously been hidden. They also served to predict the future to some degree: 'Since a picture is worth a thousand words, the aim of multimedia is to turn our old television into a kind of domestic telescope for seeing, for foreseeing (in a manner not unlike present weather-forecasting) the world that lies just around the corner' (16).

This means that a traveller who is well prepared can access a picture of what a place is going to be like before they get there. But in terms of predicting future events in any really meaningful way it is less useful. It is possible to get an accurate picture of a place that one is yet to inhabit, but this is a static representation that relies on resemblance and the likelihood that it will not have changed significantly when one actually gets there. But it tells us little of the unfolding nature of events that occur before you arrive, and nothing of the events yet to occur as repetition. Hence the premonitory characteristics of such optics are limited. And although Virilio invokes an interesting four-dimensional approach, this book wants to ask what might be at stake if we invoke the fifth dimension? Can we ever know 'the world that lies just around the corner'? To do so will mean thinking in a way that is not bound by visualization. Virilio's is a linear idea of prediction, one that lays out the future along a representable timeline. I am more interested in a non-linear universe that is unpredictable and that at best can be known in terms of the likely repetition of certain patterns.

Virilio's concentration on optics is interesting in relation to the introductory discussions around continuity and discontinuity as expounded by Bergson and Bachelard and the need for continuous discontinuity that draws on the wave particle duality that allows the discrete nature of the digital to also demonstrate the fluid characteristics of a wave, which in turn calls for the use of noise and sound as analogical devices suitable for explaining digital complexity. But his failure to fully embrace the sonic as a means of understanding digital temporality means we must draw on other sources to provide the secure ground on which a sonic methodology might rest.

What is required is a reconstituted approach to noise and sound as historical phenomena that can account for contemporary conditions. In developing such an approach, it is important to challenge the dichotomy of nature and science that too often places noise, and by association sound, in the former category. Such dichotomous and dialectical approaches regard noise and sound as natural permanent and unchanging, although technology is seen as a human and cultural phenomenon that is historical in so far as it is always changing. For Jonathan Sterne, sound takes its place in the technological realm (2003). It is a very human, cultural category that is historically contingent rather than temporally fixed. As such it is not constrained by the phenomenological privileging that leads to theological tendencies in Western thought. Such a tendency led Jaques Derrida to reverse the ontological privilege in favour of writing, a move that for Sterne was unnecessary. For Sterne, it is not a question of audio *or* visual but of placing to the fore the role and significance of sound as both a driver of history and as historically contingent. Building on Sterne's work, it may be possible to develop a mode of thinking that uses noise, sound and vibration, as organizing principles without recourse to a metaphysics of presence. That is, without falling back into the visual and to writing as both Heidegger and Derrida did.

A metaphysics of presence assumes a direct engagement with experience that is immediate, visceral and 'natural', whereas writing implores a critical detachment – a removal to a distance that affords intellectual contemplation. Sterne challenges this notion. He has carried out a detailed assessment of the historical conditions out of which sound technologies emerged. In *The Audible Past*, he plots the trajectory of sound technologies and assesses their impact. As a result of these technologies, he says: 'The voice became a little unmoored from the body, and people's ears could take them into the past or across vast distances' (2003: 1). Like in the work of Virilio, themes relating to time and space emerge out of technological development. But if Sterne invokes the audible past, this volume intends to explore the audible future by interrogating the relationship between sound and history, or more specifically between noise and temporality.

Sterne states that between 1750 and 1925, sound became an object and a domain of thought that was based on solid scientific foundation that drew on Physics, Biology and Mechanics. It emerged, he says, out of its previous idealized form in voice and music, and in doing so had a profound impact on our understanding of, and engagement with, our world. 'As there was an Enlightenment, so too there was an 'Ensoniment', he says (2003: 2). Bit by bit the technologizing of sound altered our perceptions of the world and became central to the modern project. In fact, for Sterne, sound is so historically significant in this regard that he questions the received wisdom that Western thinking is entirely orientated towards the visual, as others have suggested, saying:[2] 'But even if sight is in some ways the privileged sense in European Philosophical discourse since the Enlightenment, it is fallacious to think that sight alone or in its supposed difference from hearing explains modernity' (3).

Listening, for Sterne, can provide a different perspective, or dimension. It begets a different kind of knowledge, and as far as its role in accounting for modernity is concerned, it prefigured modern ways of seeing. Hence, he urges us to take sound seriously as a way of engaging with the world. Factors demonstrating the measurement objectification, simulation and commodification of sound in the modern period, he says, 'urge us to rethink exactly what we mean by the privilege of vision and images. To take seriously the role of sound and hearing in modern life is to trouble the visualist definition of modernity' (2003: 3). If this recommendation is appended to the thinking of Virilio as discussed earlier, then we can go further than alluding to the sonic and the temporal, and fully embrace them. Such a move has not yet occurred according to Sterne, and sound culture or sound studies has yet to impose itself on theoretical approaches in the same way that visual culture has and this is why a critical underpinning of noise, sound and music is so important.

To advance his project, and drawing on Marxist approaches, Sterne postulates the historicity of the senses. It is through sensual engagement he says that one is able to imagine a world that is other than the one we currently know. So, as will be developed further later, there is an element of not only diagnosis but of prognosis at play in his work. What it means to see and to hear, change in accordance with the prevailing social and technological conditions. So, technology in general, and sound-based technologies more specifically, emerged out of and within a complex network of cause and effect – setting agendas and responding to historical conditions. Sound technologies particularly, he says, were embedded in the modern desire to capture and make predictable the ephemeral, the strange and the magical. Voices of the dead, where the sound is separated from its source, could now be heard. Such a technological ordering of life contained within it a sense of control, and a collective confidence in the future that was a feature of modernism.

'For many of their inventors and early users, sound-reproduction technologies encapsulated a whole set of beliefs about the age and place in which they lived. Sound-reproduction technologies represented the promise of science, rationality, and industry' (Sterne 2003: 9). As such it was a very human history that did not, and still does not, account for the complex nature of a universe in which human desire to achieve certain ends is but one factor. Sound from this perspective is cited as having been central to a modernism that turns relations of space into relations of time in terms of inevitable and purposeful historical progression and advancement where recording and storing the past in the present allows us to take it with us into the future. In discussing this notion of progression, Sterne introduces the dialectic that is at play between technology and its opposite nature. By doing so in relation to sound, he is able to liberate listening within a recontextualized modernist framework. For him, the history of sound tells us something significant and hitherto underexplored about historical development. Equally, the history of sound itself is found to have developed within specific historical circumstances that did not limit it to a metaphysics of presence. He is critical of the technology–nature dualism. For him, sound is not necessarily natural and unchanging. So, he attempts to reorient it within a continuing dialectical framework. In doing so, he offers a corrective to Derrida's choice between sound and vision. Rather than simply accept the idea of sound residing in the natural realm, he places it firmly in the cultural–technological sphere, relieving it of its static nature and describing it as a mobile modern, and therefore temporal phenomenon. It takes its place alongside vision without it being regarded as a special case. In returning sound to its rightful place in history, Sterne challenges what he calls the audio–visual litany (that will be fully explored in Chapter 4). But in doing so, the culture–nature dialectic remains, leaving vibration and noise as natural remnants and sound as human, cultural and technological. And so, a new dialectic, this time between sound and noise, emerges.

Through this dialectical coming together of sound and noise, noise, for Sterne, has been technologically cloaked to such a degree as to make it almost entirely meaningless. This occurred just at the point when critical theorists – Jaques Attali in particular – were beginning to take it seriously as a critical category. But rather than supporting claims pertaining to the meaningless of noise, Sterne's work on MP3's and perceptual coding, as interpreted here, serve to reinforce the need to account for the neglected, the unwanted and the imperceptible as significant factors in any kind of critical analysis, especially one relating to digital temporality. It is essential to bring the noise back. Perceptual coding and cloaking technologies constitute a semiotic system as an economy of meaning. At their core is the acknowledgement of the claim that there is in existence a phenomena that we do not need to concern ourselves with. Digital technology can now separate the 'necessary' from the 'unnecessary' in new and efficient ways. What I am interested in, although, is the uncertain or unnecessary that is yet to be encountered. In striping away the deemed 'unnecessary', coding paradoxically reveals it or unconceals it, demonstrating a reality beyond sense that is quantifiably present, vibrating and noisy.

Sound, as defined by Sterne, becomes a very human phenomenon that is distinct from noise or vibration in so far as it becomes inexorably linked to human perception and hearing. The position being developed here proposes on the contrary that sound need not be separated from noise and vibration. Rather it should be regarded as an element of a fluid complex singularity. Sterne succeeds in establishing sound as a meaningful category, but does so at the expense of noise. In doing, so he is happy to have sound participate within a system of dialectical progression. To understand how this works, or indeed why it may not work, requires further consideration of dialectics as a mode of enquiry. More specifically it requires that we ask, what are the key characteristics of a dialectical model and why is it unsatisfactory as a means of accounting for contemporary conditions and for a critical conception of temporality in the digital age? In answering this, August Thalheimer's *Introduction to Dialectical Materialism* is a useful source of reference.[3] He says,

> The most general and the most inclusive fundamental law of dialectics from which all others are deduced is the law of the permeation of opposites. This law has a two-fold meaning: first, that all things, all processes, all concepts merge in the last analysis into an absolute unity, or, in other words, that there are no opposites, no differences which cannot ultimately be comprehended into a unity.

This corresponds directly to George Prochnick's version of noise referred to earlier in so far as all chaotic systems resolve to calm themselves.

Talheimer continues, 'Second, and just as unconditionally valid, that all things are at the same time absolutely different and absolutely or unqualifiedly opposed. This law may also be referred to as the law of the polar unity of opposites. This law applies to every single thing, to every single phenomenon, and to the world as a whole'.[4] This very clear description of the dialectical mode of thought presents the human mind as capable of infinite computation of things into unities. It is almost an algorithmical model that is capable of processing seemingly endless calculations. It can cope with even the clearest contradictions and opposites, while simultaneously being able to manage the infinite differentiation and analysis of things into opposites. From this perspective, everything is the same, and everything is different, forming a continuous discontinuity as datawave.

So far this sounds exactly like what I am attempting to describe in this book. At this stage, therefore, I appear to be conforming to dialectical logic as a cosmological model. But let us continue. Talheimer goes on to say,

> Now our question is: what is the origin of this basic law? And this is the answer: in the first place, it is a generalization of experience. In daily life and in science we constantly have to search for the identities as well as the differences of things, and experience shows that there are no rigid, fixed limits to the discovery of either. Existing limits are mobile, relative, and temporary; they are constantly being broken, reset, and rebroken.[5]

Again, I seem to be in agreement.

> Secondly, this law of the permeation of opposites may be deduced from the examination of thought itself. It is a law of thought as well as of nature. In thought this law is inherent in the basis of consciousness, and this basis consists in the fact that I know that I am a part of the universe, a part of being, and, on the other hand, in the fact that I know myself to be distinct from the external world, distinct from other things. The basic structure of thought is, from the very beginning, a polar unity of opposites, and from this all other laws of thought are derived. Furthermore, this polar unity of thought corresponds to the nature of all things.[6]

Dialectics is regarded as being rooted in nature. As such, thought is separate from, but equal to nature. It is real, essential and fundamental – (foundational as a given). It describes the way everything works and develops. Only too often it does not. Things get stuck, creating a glut, a proliferation of waste and noise. Dialectics, like digital compression, processes the excessive abundance, makes sense of it and rationalizes it. As such, it is both useful and compelling, as a natural phenomenon and a human ideal. Dialectics then works in thought as it does in nature, where process and movement are 'absolute and unlimited'. This movement is assured by the notion of negation. 'The negation of negation logically results in something positive, in thought as well as in reality. Negation and affirmation are polar concepts. Negation of the affirmation results in negation; negation of the negation equals affirmation. If I negate yes, I get no, the first negation. If I negate no, I get yes, the second negation. The result is something positive.'[7]

For Talheimer, it is the coinciding of thought and reality (as distinct dialectical entities) as a rational process that generates the new, and which is the motor of progress, and as such progress is a fundamentally human attribute even when it is regarded as natural. What results from the double negation he says is not a reestablishment of the old and the original, not simply a return to the starting point, but something new. 'The thing or the condition with which the process started is re-established on a higher plane. Through the process of double negation new qualities and a new form emerge, a form in which the original qualities are retained and enhanced'.[8] This idea is significant in terms of the critique of Henri Bergson's interpretation of dialectics that will be discussed later, which centres around the question of whether the negation destroys that which precedes, or carries something of it forward to a higher plane? It is also relevant to Jean Francois Lyotard's conception of the new, as a constant reforming of the same, without clearly demarcated beginnings or ends (Lyotard 2002), again, more about this later. Finally,

> Thesis and antithesis are dialectically united in the final proposition, the synthesis. The dialectical union must not be mistaken for the mere summation of those qualities of two opposite things which remain after mutually exclusive qualities are cancelled. Dialectical development does not occur this way; this would simply be a mixture or effacement of opposites, a hindrance to dialectical development. It is a necessary characteristic of dialectical development that it fulfill itself through negations. Without negation there is no process, no development, no emergence of the new.[9]

This kind of pure dialectics does not seem to present too many issues as far as the aims of this book are concerned. But in practice issues do arise. It is in its challenge to the separation of thought from the reality of nature that the narrative that is emerging here begins to diverge from dialectics. Tallheimer's version of a historically attuned dialectics seems to tell a story of harmony. To a degree it accounts, in its assertion of constant movement, for what is understood here as noise. Yet in the final account it is too neat, too inclined towards resolution. Despite the assertion here that noise be reappropriated as a positive term, it still needs to retain its ability to account for the unresolved, the inscrutable and the enigmatic. For Talheimer, there must be a reason to cancel mutually exclusive qualities, and this reason is self-fulfilment. But in a universe where nothing can ever really be cancelled and where the self as a subject is being rethought in relation to the real world as object, dialectics may no longer be a sufficient means of accounting for our contemporary mode of existence.

Another key problem with dialectics is that it is teleological. It posits specific ends, as self-fulfilment, that arise from conscious recognition of, or conscious imposition on, our existence, whether material or ideal. It is a systematic way of imposing some kind of order on chaos, without which we are left, according to its advocates to the indiscriminate whim of history, or even worse nature (even when the mastery of nature is regarded itself as natural). But ultimately it cannot account for the indeterminacy of events and their interactions and as such needs to be amended, and modified, if not rejected altogether.

Michel Foucault was a central figure in rejecting dialectics and an advocate of a move towards a mode of engagement that foregrounded a much more nuanced version of power–resistance based on discourse (Foucault 2004). Dialectics for him proposed a mode of thought where certain conditions and their opposites are not mutually exclusive but enter into dialogue – each vindicating the other until resolution emerges as a new set of conditions. For Foucault, this situation could be explained as less of a coming together of opposing positions and more of a constant multifaceted reality or discourse where power and resistance, as opposites sustain each other in a constant and unresolvable dance that is akin to Hainge's account of noise. Yet according to some commentators, his argument retained certain features of dialectics.

'If Foucault is compatible, in specific ways, with a sophisticated dialectics that is Hegelian-inspired, it would go some way to constructing the basis for a more productive engagement in the future between, for example, Foucauldians and Deleuzians, on the one hand, and critical theorists (perhaps Žižek and Jameson most importantly) on the other'. (Grant 2010: 221)

But for Foucault himself, 'it was necessary to free ourselves from Hegel – from the opposition of predicates, from contradiction and negation, from all of dialectics' (Foucault 1977: 186 in Grant 2010: 222). The methodical nature of dialectics was too stringent for Foucault and could not account for the complexities of intertwining genealogies and ruptures that he identified as features of the interplay of power and resistance – a relationship understood here through the figure of noise. Noise understood in this way is not something to be silenced but something to be amplified, celebrated and cherished, taken as a feature of difference that cannot be rationalized. Power is noise, resistance to power is noise; noise is the presence of everything not yet manifested as anything. The interesting challenge from the point of view of this study then is how to categorize or think through activity and situations that are not dialectically oppositional? Is it possible to meaningfully account for a situation where irresolvable incompatibilities are a constant feature, as they currently seem to be, rather than being regarded as essential systematic features of each other combined through coincidence, and ultimately resolvable? This raises the distinction that John Grant highlights between contradictions and antagonisms:

Quoting Adorno:

> Dialectical contradiction is experienced in the experience of society. Hegel's own construction, formulated in terms of the philosophy of identity, requires that contradiction be grasped as much from the side of the object as from the side of the subject; it is in the dialectical contradiction that there crystallizes a concept of experience that points beyond absolute idealism. It is the concept of antagonistic totality. (1993: 78)

Grant goes on to say,

> Even when Hegel's dialectical philosophy is revealed as inadequate [by Adorno], its richness illuminates what Foucault did not believe: that the trials of consciousness implicate those of an entire society [critical thinking as capable of producing universal solutions to complex situations]. These trials are experienced not as contradictions, however, but as antagonisms. Antagonism articulates a sense of opposition, hostility, and even suffering, which contradiction alone often cannot convey. (Grant 2010: 225)

This describes Adorno's attempts to resolve the inability of ideal forms to overcome objective actualities. It is a more complex account that takes seriously the existence of irresolvable states. As such it may better describe our contemporary state of affairs. The increasingly complex nature of the system could though still be systematically critiqued for Adorno. He would go on to explain circumstances that did not fit with formal dialectical accounts. For Adorno, such antagonisms were present in the pattern of bourgeois capitalism that had established itself to such an extent that even the clearest opposition to it was subsumed and integrated into that pattern. To enact a shift or a disruption, a new pattern had to be initiated. He found the basis for such a patterning in music generally and in Schoenberg's atonal system more specifically.

If Hegel, Adorno and Foucault all faced problems, what can I do to counter them through the figure of noise? Might the answer lie in music? For Adorno, it was certainly a significant realm for potential enquiry, and if properly understood in relation to noise, as it is here, it may reveal significant new forms of knowledge not accessible through visual means alone. To be meaningful although, noise, sound, music and listening need to be enacted in a very specific way, and this is the challenge that lies ahead.

Much of what is discussed here does not fit with a neat dialectical model. Rather it is characterized by alternative and antagonistic rather than oppositional strategies; sideways or even backwards moves that invoke a multi-temporal dimensionality. But does that render the future unpredictable? If so then many of the positions explored in this book that declare sound and/or music to be heraldic, will need to be challenged. Indeed, the very notion of 'future' might need to be reconsidered. And this will have implications for the ways in which we think about time, organize it, or submit to its ultimate will. Dialectics as a method of temporal engagement serves to constitute an imposed order of linearity. The chaotic realm however, does not conform to such an order. As such it is without conventional time, in so far as time is taken to be composed of both measure and/or experience. But if time is understood instead as a non-linear matrix of decay and infinite reformation, then things may be different. Time from this perspective could operate at multiple rates in multiple modes simultaneously, and without the need for subjective validation. This is time in the age of the digital (Barker 2012).

So, if dialectical methods are inappropriate for understanding this reconstituted multi-temporality, then there is a need to develop a more suitable mode of enquiry that can tolerate inconsistencies and accept the multiple counter-intuitive anomalies that the contemporary world demonstrates. To understand such a world, and the science that underpins it, requires an approach that interrogates the world not as fixed and visual but as mobile and sonorous. To support this, the work of Wilfrid Sellars can be of considerable use.

Sellars proposed a union of subject and object through his discussion of the *manifest* and *scientific image* that will be read here through the figure of noise. He was a key thinker, along with Willard Van Orman Quine and Ludwig Wittgenstein in terms of his critique of sense data empiricism and his challenge to what he called the 'Myth of the Given'. Sellars, like Kant before him, was keen to resolve the bifurcation that faced him: In this case, the division between rationalism and empiricism, or what the mind does *with* data and what is given *as* data. Either empiricism accounted for everything, he postulated, or rationalism did. Yet neither seemed to suffice. Instead, he explored the possibility of entwining the two as part of a singular system. As he proceeds bifurcations are constantly getting resolved in his work, only to reoccur as features of a system that can be constituted of dualisms and singularities simultaneously.

Like Wittgenstein, Sellars based his analytical system on language and on what he called psychological nominalism where concepts and knowing are rooted in words and language that are in turn social systems.

> The essential point is that in characterising an episode or a state as that of knowing, we are not giving an empirical description of that episode or state; we are placing it in the logical space of reasons, of justifying and being able to justify what one says. In other words, knowledge is inseparable from a social practice – the practice of justifying one's assertions to one's fellow-humans. It is not presupposed by this practice, but comes into being along with it. (Sellars 1997: 4)

Knowledge then is not an independent state of being that language merely describes. Language is not a foundational analytical principle that is brought about by empirical experience in the Humean sense, rather language as social practice *accompanies* the coming into being of episodes and states. The act of characterizing and the episodes and states being characterized are not the same thing, but form a dynamic interaction (as noise) that creates what reality is, again in a similar manner to the resistance in an electromagnetic field as described by Greg Hainge. This is not a really real and a perception of a mediated real, but an actual real that is not founded on a single premise (neither entirely rational nor empirical). So, knowledge emerges as episodes and states conjoin with discourse as shared experience expressed in language to create the real. That is reality, the process of conjoining where language is constituted as 'sonic matter' and where humans constitute 'organic support' in a non-linear and chaotic orchestration (Delanda 1997).

Sellars refuted the idea that episodes and states required a linguistic injection of life, or that language was an afterthought in terms of having emerged out of an encounter with a thing, episode or state. The temporal sequence of cause and effect has no place here. This is rather a temporality of noise – an everything playing out all at once (Brassiere 2007), and not in a logical order. The world according to Sellars' account cannot be reduced to foundational principles, but rather is a set of socially infused standards and agreements that constitute truth and knowledge as part of a dynamic system. Sellars challenged the distinction between 'what is given to the mind' and what is 'added' by the mind. Richard Rorty in his introduction to *Empiricism & The Philosophy of Mind* (1997) highlighted the following often-quoted section of the book: 'Empirical knowledge, like its sophisticated extension, science, is rational, not because it has a foundation but because it is a self-correcting enterprise which can put any claim in jeopardy, though not all at once'. He went on to add, 'This sentence suggests that rationality is a matter not of obedience to standards (which epistemologists might hope to codify), but rather of give-and-take participation in a cooperative social project' (Sellars 1997: 6).

One important question that arises in relation to the task at hand here is: Is it possible for such give and take participation to be codified organized and meaningfully represented? Or is it too complex to be accommodated within any system – either dialectical or digital? If so we may be tempted to submit to the unpredictable and precarious nature of reality. This is of course not the intention here.

Sellars cited a system in which language co-evolved, not as an abstract phenomenon peculiar to man but as a faculty that arose in tandem (not as linear succession) with the circumstances of its inception. Such a system means there is no need for an either/or, reason or empiricism, debate. Their interaction is a process that takes place in the movement between the rational and the empirical. As a means of extrapolating this entanglement of subject and object, Sellars drew on a myth of his own making. He created what he called 'The Myth of Jones' – a story that was intended to test the direction of flow between mind and language, and which explained why it was possible to accept Wittgenstein's doubts about what Sellars called 'self authenticating non-verbal episodes' without sharing Ryle's doubts about the existence of such mental entities as thought and sense impressions (Sellars 1997: 6). It can be accepted within a Sellarsian framework, that sense impressions do exist, but they only become significant once verbalized through language. So, for Sellars, it was possible to share Wittgenstein's doubts without questioning the *existence* of mental entities entirely. Mental entities exist but not as non-verbal entities where each individual is the author of a unique reality. Instead mental entities and sense impression pre-exist language as noise and resistance, and they gradually interact to form patterns from shared/common experience that co-evolve through language into thought.

> Sellars treatment of the distinction between mind and body has been followed up by many philosophers of mind in subsequent decades. He may have been the first philosopher to insist that we see 'mind' as a sort of hypostatization of language. He argued that the intentionality of beliefs is a reflection of the intentionality of sentences, rather than conversely. This reversal makes it possible to understand mind as gradually entering the universe by and through the gradual development of language, as part of a naturalistically explicable evolutionary process, rather than seeing language as the outward manifestation of something inward and mysterious which humans have and animals lack. As Sellars sees it, if you can explain how the social practices we call 'using language' came into existence, you have already explained all that needs to be explained about the relation between mind and world. (Rorty in Sellars 1997: 8)[10]

This describes a world where thought and reality are not separated. This relationship can be set against the temporality of noise – like a signal that is scrambled – a vibrating chaotic system where everything relates to everything and everything is possible, and out of which patterns form driven by cosmic and quantum forces (that are in some sense atemporal); without linearity or sensible notions of beginning and end or cause and effect. In making this assertion, Sellars was attempting to extend the Kantian synthesis, and his contemporary, the analytical philosopher Robert Brandom, was keen to extend it even further into a Hegelian realm where the (Gordian) knot that ties subject and object is increasingly tightened. Brandom suggested that Kant's contribution was to successfully counter the ideas of British empiricists, who doubted all knowledge until the unmediated link between subject and object could be proved. 'That habit is characteristic of philosophers who, in Brandom's terms, are "representationalist" (like Descartes and Locke) rather that "inferentialist" (like Leibniz, Kant, Frege, the later Wittgenstein and Sellars). The former take concepts to be representations (or putative representations) of reality rather than, as Kant did, rules which specify how something is to be done' (Sellars 1997: 9).

These rules are central to the ideas being discussed in this book with its focus on pattern formation/deformation as information and the temporality of noise. Because noise is by its nature not a stable state, it is necessary to acknowledge, and as best as possible understand, how it resolves to produce instances of predictability and stability, even if they are always only temporary. For Brandom, as it had been for Sellars: 'Following out this side of Kant's thought, rather than the side which led him to the sceptical conclusion that we could have no knowledge of things as they are in themselves means emphasizing the passages in Kant which anticipate, Marx, Dewey, and Habermas, as opposed to those which connect Kant with his predecessors' (Sellars 1997: 9).[11] The problem then becomes one of figuring out the rules of how states of disarray, confusion and noise are mediated to form those aspects of our universe that can be recognized as ordered, and to do so without recourse to representation. For Hegel, this problem of mediation had been not so much of a problem as a solution. By placing it as central to his dialectical system, he made it part of what knowledge was rather than simply a route to or from knowledge – as did Marx. What remained for both however was the primacy of the human, as it did ultimately for Heidegger, and this meant that for contemporary thinkers, particularly those associated with Speculative Realism, the fundamental problem was never really resolved and remained tethered to a correlationist anchor (about which more later).[12] All that can be known for them is a very particular human world that has been ordered through language to make sense.

Sellars contribution to this problem was to challenge the 'Myth of the Given'. For him knowledge was not derived from a procedural flow of information from object to subject. Knowledge of what is perceived cannot be abstracted from the social and conceptual processes that result in perception. The 'social' itself forms out of noise and as it constitutes itself it makes noise – noise is the sound of the social forming. It is not a case of reason (or pure reason) being able to access this precognitive realm or not, but rather that cognition is formed of precognitive material that endures in cognition. It does not separate into mind–body or subject–object. This notion drew Sellars' towards a rejection of the 'arche', a rejection that has implications for the ideas proposed by Quentin Meillassoux as will be discussed later.

Returning to 'The Myth of Jones', Sellars writes,

> I have used a myth [of Jones] to kill a myth – the myth of the given. But is my myth really a myth? Or does the reader not recognize Jones as Man himself in the middle of his journey from the grunts and groans of the cave to the subtle and polydimensional discourse of the drawing room, the laboratory, and the study, the language of Henry and William James, of Einstein and of the philosophers who, in their efforts to break out of discourse to an arche beyond discourse, have provided the most curious dimension of all? (Sellars 1997: 10)

This dimension that Sellars says philosophers are endeavouring to escape to is one in which human beings play no particular part. It is an originary space and time from which life emerges. Yet it is only certain emergent life forms that can reflect on it, and they do so through language and representation, and so the problem begins over, and over again. But how does this relate to the temporality of noise? As a temporal dimension noise includes an economy of relations among all aspects of the universe and is not restricted to human experience or rational thought. It describes the co-evolution of conditions through a discourse among material forces that *become* linguistic rather than existing as source material for language.

The crucial question for this volume is how the journey from grunts and groans to polydimensional discourse sits within a temporality of noise? In relation to this Manuel Delanda says:

> Human languages are defined by sounds, words and grammatical constructions that slowly accumulate in a given community over centuries. These cultural materials do not accumulate randomly but rather enter into systematic relationships with one another, as well as with the human beings who serve as their organic support. The 'sonic matter' of a given language (the phonemes of French or English, for instance) is not only structured internally, forming a system of vowels and consonants in which a change in one element affects every other one, but also socioeconomically: sounds accumulate in a society following class or caste divisions, and together with dress and diet, form an integral part of the system of traits which differentiates social strata. (Delanda 1997: 184)

This describes a 'sonic economy', and relates directly to Sellars' proposal that language is social and cooperative rather than a response to a pre-existing given. But how it came into being has still not been established.

Sellars wrote in his *In the Space of Reasons* (2007): 'The conclusion is difficult to avoid that the transition from pre-conceptual patterns of behaviour to conceptual thinking was a holistic one, a jump to a level of awareness which is irreducibly new, a jump that was the coming into being of man' (Sellars 2007: 374). This 'irreducably new' situation brought with it new beginnings and this raises the question of whether the 'jump to a new level of awareness' was a continuity or discontinuity? It requires that we consider a situation where everything was/is going along in a particular direction until an unforeseen break occurred and man suddenly appeared. This idea of the new as a radical break or rupture needs further discussion, and it is worth pausing momentarily to think through some of its features and implications.

In his book *Without Criteria*, Steven Shaviro revisits the notion of the new. He imagines a world where not Heidegger but Alfred North Whitehead dominates, and where we ask not why there is something rather than nothing, but how it is that there is always something new? Both Heidegger and

Whitehead invoke temporality as a central thread in their work, but do so in different ways. Shaviro writes,

> *The question of beginnings* Where does one start in philosophy? Heidegger asks the question of being: 'Why is there something rather than nothing?' But Whitehead is splendidly indifferent to this question. He asks instead: 'How is it that there is always something new? Whitehead doesn't see any point in returning to our ultimate beginnings. He is interested in creation rather than rectification, Becoming rather than Being, the new rather than the immemorially old. I would suggest that, in a world where everything from music to DNA is continually being sampled and recombined and where the shelf life of an idea, no less than of a fashion in clothing, can be measured in months if not weeks, Whitehead's question is the truly urgent one. Heidegger flees the challenges of the present in horror. Whitehead urges us to work with these challenges, to negotiate them. How, he asks, can our culture's incessant repetition and recycling nonetheless issue forth in something genuinely new and different. (Shaviro 2012: x)

Heidegger's desire to flee the present is an interesting one. His claim to a higher authority was based on a *way* of thinking that could escape the historical conditions of inherited thought. It was an attempt to step out of time, yet to do so via recourse *to* time as a means of denouncing essentialism in favour of a methodology that ultimately instantiated another kind of essentialism. Heidegger's intention was to question the views of the humanist tradition, the idea that human beings possessed essential qualities that allowed them to comprehend their world and represent it in a rational manner. Rather human beings were/are historically contingent and operated as an 'effect of language' (Shaviro 2012: xii). This questioning of essentialist subjectivity was contrasted with the proposition that non-essential humans were the 'site where language manifests itself' as Dasein, that particularly human way of being, that when extricated from its representational shackles, could uncover a different truer essence.

How though does this fit with the Sellarsian model of language as a product of evolution that is manifested in man and not as an essential feature of man? Sellars was interested in both the moment when man emerged from noise into language and also, like Whitehead, in the constant becoming that the manifest and scientific images provoked. Man was not the effect of language as he was for Heidegger.

> Nothing could be more foreign to Whitehead than this whole polemic. As before, this is not because Whitehead is concerned to defend what Heidegger is attacking, but because his interests lie elsewhere. Whitehead does not see the subject as an effect of language. Rather he sees subjectivity as embedded in the world. The subject is an irreducible part of the universe, of the way things happen. There is nothing outside of experience; and experience always happens to some subject or another. (Shaviro 2012: xii)

This approach seems to resonate with that of Sellars. Of course, for Whitehead, the subject need not necessarily be human. The subject, as both object and as sense receptive being (a process that includes thought) is constituted as pattern arising out of chaos, or as harmony out of noise, *as noise*, or another kind of noise. It is always moving as a flow of resistance, always changing, always renewing where 'new' is not necessarily originary but is rather an expression of difference.

Returning to Sellars, in his book *In the Space of Reasons*, Sellars set out the categories *manifest image* and *scientific image*. It is worth exploring these predominantly visual categories within the sonic context of this book. Sellars stressed the temporal nature of the manifest image as having a 'quasi historical dimension'. There was for Sellars a 'moment' when man recognized himself as man, and from that point on recognition and consciousness of that recognition become ever more entwined. The manifest image is an originary 'original image', a moment, and also a continuing process of refinement (2007: 375). The process of refinement was both empirical and categorical. In so far as it involved thought it also involved standards of measurement and accuracy relating to that thought. As such it (the manifest image) might be said to be scientific in nature and approach. But this is not what Sellars meant by scientific in relation to the 'scientific image'. The scientific image might better be described he says as a 'postulational' or 'theoretical image' (375). Although they are demarcated as distinct, this dualism must be understood within the context of Sellars's Kantian ambition to unite the two poles.

The dual aspects of the manifest image – the moment of the 'original image' which describes the 'framework in terms of which man came to be aware of himself as man-in-the-world' contains a paradox: 'the paradox of man's encounter with himself . . . Man couldn't be man until he encountered himself' (Sellars 2007: 374). The paradox states that man pre-exists, but cannot exist *as* man until he encounters himself. This implies a 'scientific' existence that predates a manifest existence, which is of a different kind – they are two aspects of the same thing in a dialectical bind. Sellars says,

> There is a profound truth in this conception of a radical difference in level between man and his precursors. The attempt to understand this difference turns out to be part and parcel of the attempt to encompass in one view the two images of man in the world which I have set out to describe. For as we shall see, this difference in level appears as an irreducible discontinuity in the *manifest* image, but as, in a sense requiring careful analysis, a reducible difference in the scientific image. (Sellars 2007: 374)

The complexity involved in understanding Sellars' position in relation to 'The Myth of the Given' and the attendant rejection of dualisms – although done in a way that uses a unifying *of* dualisms – might be better explained via recourse to sound; stereophonic rather than stereoscopic. When read through the figure of noise, Sellars is describing the movement from prehuman to the human and to what is now called post-human as a continuous discontinuity where the radical difference between noise and sound is reducible ultimately to the operation of noise itself as a vibrating totality out of which pattern emerges. It is also describing the birth of critical distance, a moment where we step out of noise and into the self-identifying image. But what if we do not step out, or if we step out always to return, to reclaim noise and engage with it once more?

The originary moment that Sellars was at pains to identify was present as a sonic event, as a primal scream. It was a scream that became refined through language as aural, proceeding to the visual, eventually to return to the sonic as invisible phenomena came to prevail in science. In accounting for this process, Sellars described a stereoscopic system where one lens (the manifest image) had traditionally dominated, creating a tension between man as he saw himself and as he really was. Sellars' stated aim was to develop his argument in such a way as to explain how the manifest image and the scientific image could resolve the bias and blend in perfect stereoscopy. To explain this, he invoked the ideas of Spinoza. Spinoza had contrasted man as he thought himself to be and man as he scientifically discovered himself to be. The two positions were at odds with each other in a way that Sellars sought to resolve: 'But if in Spinoza's account, the scientific image, as he interprets it, dominates the stereoscopic view (the manifest image appearing as a tracery of explainable error), the very fact that I use the analogy of stereoscopic vision implies that as I see it the manifest image is not overwhelmed by the synthesis' (Sellars 2007: 377).

By advocating synthesis, Sellars was conforming to the kind of dialectical method described earlier, where nothing is lost in the process of negation, at least nothing useful. The scientific image was in effect liberated from abstraction.

In terms of the scientific image he said,

> The scientific image of man-in-the-world is, of course as much an idealization as the manifest image – even more so, as it is still in the process of coming to be. It will be remembered that the contrast I have in mind is not that between an unscientific conception of man-in-the-world and a scientific one, but between that conception which limits itself to what correlational techniques can tell us about perceptible and introspectible events and that which postulates imperceptible object and events for the purpose of explaining correlations among perceptibles. (Sellars 2007: 387)

The inclusion of the imperceptible is crucial, and I would suggest that a sonic analogy would better serve the purpose that Sellars intended.

In describing the coming together of the manifest and the scientific image Sellars gives a detailed account of previous attempts at synthesis, saying:

> The same considerations which led philosophers to deny the reality of perceptible things led them to a dualistic theory of man. For if the human body is a system of particles, the body cannot be the subject of thinking and feeling, *unless thinking and feeling are capable of interpretation as complex interactions of physical particles;* unless, that is to say, the manifest framework of man as one being, a *person* capable of doing radically different kinds of things can be replaced without loss of descriptive and explanatory power by a postulational image in which he is a complex of physical particles, and

all his activities a matter of the particles changing in state and relationship. (Sellars 2007: 397)

State of mind then becomes a state of matter (quantum matter) where the presence of an observing subject as material object can have a discernible impact on the laws of nature, as famously demonstrated by Richard Feynman's double slit experiment that demonstrated the presence of a wave–particle duality.[13]

Sellars' account seems to be an apt description of reality, and one that is shared with Whitehead, and further echoed here. The perception of particles by particles forms patterns and relationships where noise is refined into language through resistance, ultimately to return to noise as it reaches its limits. This process moves through the interplay between the manifest and scientific generating a charged energy source. As a historical trajectory, or a temporal shift, this is interesting in terms of the move from acoustic to visual space, and the emergence of figure from ground, as described by Marshall McLuhan (1988). For Sellars, the ideas of original recognition and refinement are visual concepts. They tell a story of discontinuity, of a sudden rupture – a moment of coming into being that is represented on the visual plane as image. It is a brilliant account of knowledge and its materiality as an entwining subject–object bond where the manifest is one aspect of a stereoscopic view that also includes the scientific image. However, the scientific element in so far as it remains postulational and theoretical is an abstract image – the ground from which figures emerge. It is also continuous in nature. It is that which makes everything possible but which in itself is meaningless (the virtual). But what if it was meaningless only in terms of visual representation? What if as noise it was infinitely meaningful – stereophonic rather than stereoscopic?[14]

In order to make any kind of advance from the Sellarsian stereoscopic model, although, it is important to make the argument for a (stereo)phonic reading of his work – one that is capable of accounting for the 'acoustic' as an enhanced multisensory realm.[15] To this end it is worthwhile examining other similar interventions where enhancements to existing theoretical models have been argued for. In the first instance in relation to the work of Merleau-Ponty and *his* challenge to dualisms – so much a feature of this book. Ponty, like Sellars, was keen to resolve the Cartesean framework, in this case via a celebration of embodiment. Although his work dealt primarily with the visual, there are moments in his thinking, it is claimed, when sound is important. In *Vibrating Colors and Silent Bodies. Music, Sound and Silence in Maurice-Merleau-Ponty's Critique of Dualism* (2012), Amy Cimini writes: 'Even though, as I will explain, Merleau-Ponty has notably little to say about music as such in his voluminous oevre, his understanding of the body's centrality to social life and the production of knowledge speak powerfully to music studies' sustained interest in embodied knowledge in performance and listening' (Cimini 2012: 354).

Merleau-Ponty made the point that we are immersed in our world, surrounded on all sides and not simply inclined towards what we see in front of us. So, as Cimini points out, the fact that we are 'surrounded by the world' points to a sense of immersion historically ascribed to hearing and listening rather than seeing and looking. In exploring this, Cimmini refers to the work of Frances Dyson and her interpretation of Merleau-Ponty from a sonic perspective saying: 'Dyson's reading is extremely compelling for a musically orientated engagement with Merleau-Ponty's philosophy. If, Merleau-Ponty's conception of being is, in fact, sonor-ous or vibrational, then listening should give us special purchase on – or privileged access to being' (356).

Cimini, though goes on to challenge some of Dyson's assumptions about listening, specifically the idea that it is non-directional:

> Dyson's reading subscribes to a historically normative conception of listening as immersive, non-agential and non-directional. Indeed, while this construal can be illuminating, it often obscures more detailed and nuanced ways of understanding the epistemological and ethical implications of the ear's openness. The second challenge is specific to Merleau-Ponty's philosophy. Although Dyson asserts that the Merleau-Pontian subject is so open to the vibrational movement of being that she 'might as well be a listener' (Dyson 2009: 120), Merleau-Ponty is overwhelmingly concerned not with sound but with vision throughout his philosophical career. (Cimini 2012: 356)

More nuanced approaches might take account of directional personalized sound, as will be discussed later with reference to Kroker and Kroker's essay *Codedrift*, and may not leap to generalized immersive conclusions about sound, but in so far as it operates in a 360-degree arena rather than on a 180-degree plane, it can be usefully contrasted with the visual. So, taking as a starting point, Dyson's 'provocative injunction' to think of subjects as 'listeners', it is possible to map Merleau-Ponty's concepts onto notions of sonic vibration that 'define sound through its refusal to respect material or conceptual distinctions between subject and object, human and non-human, interior and exterior' (Cimini 2012: 360). This refusal then allows, as was the case above with Sellars, subjects to be understood as being patterned from interacting particles.

Cimini's cautionary challenges are valid, and to avoid them being applied to the work being presented here, listening and sound are replaced with noise as a central figure so as to circumvent the simple replacement of viewers with listeners. For in noise there may be a tendency towards pure sensation as described by Merleau-Ponty himself:

> The greyness which, when I close my eyes, surrounds me leaving no distance between me and it, the sounds that encroach on my drowsiness and 'hum in my head' perhaps give some indication of what pure sensation might be. I might be said to have sense-experience (*senir*) precisely to the extent that I coincide with the sensed, that the latter ceases to have any place in the objective world, and that it signifies nothing for me. (Ponty 2007: 3)

Merleau-Ponty was not advocating such pure sensation as such, but a much closer investigation of perception where the body is inescapably implicated in both rational and empirical activity. He was not interested in denying or ignoring the 'mental' aspects of life but wanted to suggest that the use of 'mind' was inseparable from the body which was situated and physical in nature. 'This means simply that the perceiving mind is an incarnated body, or to put the problem in another way, he enriches the concept of the body to allow it to both think and perceive. It is also for these reasons that we are best served by referring to the individual as not simply a body, but as a body-subject' (http://www.iep.utm.edu/merleau/). I, on the other hand, am more inclined towards the greyness where nothing is signified – other than the constant renewal of signification itself as an event or intense pattern. And this relates directly to Wilfrid Sellars' position as outlined earlier where a state of mind becomes a state of matter.[16]

The idea that 'pure sensation' is somehow invisible gives rise to an emphasis on sound as affective rather than intellectual. It is an emphasis that was employed by the artist Paul Klee in his attempt to forge an art practice wherein painting could attain the inherent quality of music. But this was another example of enhancing a primarily visual theoretical model with the spirit of sound. A more detailed discussion of Klee's contribution to this debate will take place later, but for now suffice it to say he was unimpressed by overly intellectual modern music.

Klee had been keen to extricate himself from the shackles of 'academic painting', and to embrace the musical spirit of Mozart to realize his own artistic ambitions. He also felt that music, although, as the purest of art forms, had gone as far as it could and that painting was its logical successor. This belief marked an interesting temporal juncture between music and painting for Klee – a juncture characterized by ends and beginnings. For Klee, although music had gone as far as it could, there was still something about the sonic that could inflect painting and empower it to even greater heights. This inflection was essentially dialectic in character, in so far as something of the old would need to be present in the new. His rejection of music as a form of expression that had become too academicized however continues to be an interesting one. It relates to the manifest and scientific image of Sellars and to the pure sensation of Merleau-Ponty. It also implies that, if performed appropriately, music specifically, and sound and noise more generally, belong to a spiritual realm that is not conducive to scientific investigation – that it is a pure sensation, affective and emotional. This idea is vehemently opposed here, and is replaced by an approach to noise, sound and music that is immanent – neither manifest, or scientific, wave or particle, but as always both simultaneously.

Much of the argument as it has developed so far then has been orientated towards the need for sound to be taken more seriously. But as a temporal art (but possibly not uniquely so) it must also embrace the spatial and the visual (as aspects of noise) as mobile multidimensional phenomena that can extricate themselves from the philosophical stasis that has dogged them for centuries. To develop this logic further, it is worth considering the contribution made by Christopher L. Witmore. He argues that the historical neglect of sound must be addressed, and does so, interestingly, from an archaeological perspective. He argues that

> while the visual is linked with spatial properties that are resistant to change, the aural is connected with the temporal and is considered momentary and fleeting in nature. Still, it is argued that sound as a quality of things is fundamental to human sensation – to being. In building upon a non-modernist notion of time where entities and events quite distant in a linear temporality are proximate through their simultaneous entanglement and percolation I suggest we might learn what we can understand from tuning into the acoustic properties of the material past. But rather than reproduce an unnecessary dualism between seeing and hearing, this endeavour will require us to relearn how to see and hear at the same time through other, complimentary modes of articulation and engagement. (2006: 267)

Witmore alludes to a sonic portal to the past in contrast to Attali's heraldic portent of the future (and both do so by invoking the noise that is present in the paintings of Pieter Bruegel the Elder). In doing so he prompts some interesting discussion on the dialectical relationship between time and space. Like Salome Voegelin's notion of timespace, this is yet another attempt to collapse a dualism, in this case an attempt to merge the audible and the visual (Voegelin 2010). In a similar vein to Sellars' stereoscopic synthesis, it is directed towards collapsing and eradicating the dualism altogether or entwining the dual aspects so tightly as to make them indistinguishable.

For Witmore 'noise connects us to deeper textures of the material world and qualities of corporeal experience'. To hear noise, he says, is to hear things. And as noise is resistant to the flow of time, it is uniquely equipped to allow us access to the past. Referring to Michel Serres, he gives the example of sea noise that 'never ceases; it is limit-less, continuous, unending, unchanging' (Serres 1995: 13). Such sounds may contain information that spatially located visual methods cannot unearth. Yet these background noises are consistently filtered out in archaeological practice. (Serres 1995: 15; see also Witmore 2006). Such filtering implies that a degree of unreliability is inherent in the temporal, and also, therefore in noise and sound. But for Witmore, as for me this is not the case.

> Imagine the *belles noiseuses* of a lively countryside during harvest as depicted in Pieter Bruegel the Elder's *The Harvesters* (Figure 2). Painted in 1565, *The Harvesters* portrays field hands reaping corn in the countryside in the month of August. For Tim Ingold, the painting 'vividly captures a sense of the temporality of landscape' (1993: 164). In his well-known discussion of the landscape's temporality, Ingold focuses on six aspects of the bucolic scene that unfolds before the viewer's eyes, 'the hills and valley, the paths and tracks, the tree, the corn, the church, and the people' (1993: 166). Each of these elements has a different temporal rhythm; each has a different pace within the flow of time over the long, intermediate, or short term. For Ingold, this temporality is what forms the subject matter of archaeological inquiry. (Witmore 2006: 274)

Here the act of looking also has a sonic dimension. The passage describes listening to the past as an appropriate archaeological practice, a listening to sound that is otherwise locked in. It is a listening that is similarly practiced by the dream reader in Haruki Murakami's *Hard Boiled Wonderland and the End of the World*, in which he describes objects that would not give up their secrets until they are tuned into, whereupon they vibrate and hum until they emit light that can be read as dreams from the past in a land that is atemporal.

In this novel, Murakami gives an account of two worlds that are inextricably linked – the world of linear time, and a parallel world where time ceases to pass, and into which the main character is guided by a girl rendered incapable of speech by her grandfather's sound removal technology. In the latter world, time is locked in the skeletal skulls of strange creatures and their secrets can only be revealed through an intense and meditative form of listening. This is the world of the known and the unknown, the manifest and the scientific. For Murakami's central character, it is a race against time to stay in one world and escape slipping into the other. But whether it was such a simple either/or question is debatable? Murakami wrote the story in stereo. In doing so he was able to describe a single mind that existed in two dimensions – dual aspects of a singularity without one cancelling out the other. But in the end, one prevails and the reader is left to ponder a situation where some of the story's characters live in one realm, some in the other. They exist in parallel universes, side by side, but seldom do they merge. Witmore, like Sellars, would endeavour to force such a merger.

For Witmore sound has long been neglected in relation to the visual, and this neglect he says provides a catalyst for raising questions pertaining to a number of other classic divides: between humans and things, space and time, past and present – and for the proposal of some potential alternatives. He argues that the failure to account for acoustic data within any historical picture is in part linked to the instruments and media used in the process of analysis. Rather than treating human beings as autonomous freestanding subjects who intention towards the world as autonomous and critical beings, he suggests that maps, along with theodolites, tapes and compasses (not to mention photography and cameras), have had a role to play by 'enhancing and redirecting our senses'.

> In archaeological practice (with the help of the brigades of visual media, instruments, and our knowledge of the visual perspective) sound is often relegated to the chaotic background agitation of the material world. The background noise – too chaotic, too confusing, too multiple, too messy – is temporally situated. Whether it exists as the tick of the clock, the ring of the bell, the buzz of a passing bee, the clash of thunder, or the steady rhythm of stiletto heels across the pavement of the Via Sacra cutting through the Roman forum, sound is momentary and fleeting. Sound is transient. (Witmore 2012: 272)

It is precisely this chaotic noise that is under investigation here. It is a noise that operates against the grain of modernist thought that neatly separates out the past and the present, assigning them to specific temporal categories that can be named and measured, and that have convenient beginnings and ends. Time in the modernist sense is compartmentalized and we become separated from our premodern predecessors by 'Copernican revolutions, epistemological breaks and epistemic ruptures so radical that nothing of the past survives in them – nothing of that past ought to survive in them' as Witmore says making reference to Bruno Latour (1993: 68). Yet again the conundrum of historical remnants emerges. In dialectical terms, such revolutions elevate historical modes to a higher plane, as time's arrow proceeds. 'However, the notion of linear flow, as Michel Serres reminds us, does not describe the nature of time itself; it is rather only one form of temporality. Time itself is much more complex. Time is much more chaotic' (Witmore 2012: 278). And in such chaotic modes of time something of the past does survive. To show how, Witmore refers to the work of Laurent Olivier who extends the existential mode to include a range of non-human agents. He is interested in the materiality of the past and its folding into the fabric of the present, where we can continue to engage with it as particles interact with particles. Such a 'folded, chiasmic and entangled time' is for Witmore profoundly archaeological in so far as to experience the recurrence of noise is to hear the past. His is a multi-temporal version of history that draws simultaneously from the 'obsolete, the contemporary and the futuristic' (Serres with Latour, 1995: 60), and as such serves to support a conceptualization of history as a temporality of noise wherein subject and object, past and future fuse in the present.

Like Murakami, Witmore instantiates the sounds of the past, as living and significant, into a folded version of time which is non-linear. In this non-linear, non-dialectical model of temporality–history, noise can produce an echo of the future. *Whatever can happen already has happened* (i.e. if we step out of human time, and into a material time of virtuality and potential, where the components of the future are already in existence, just as they were in the past) and some of it rebounds into the present where all that is possible is already accessible to us as noise and vibration.

Despite placing a significant emphasis on noise and sound, Witmore is acutely aware of the dangers inherent in simply having the audible replace the visual. He cites Wolfgang Welsch's *Undoing Aesthetics* (1997), in which he says the author conflates Western philosophy and the dominance of the visual in modern culture. To counter centuries of wrongdoing in the form of colonialism and racism, Welsch proposes an auditive turn. But he does so without any critical explanation of how replacing one dominant sense with another might accomplish such a task. For Welsch both hearing and vision were 'long-range senses', the difference between them being that vision was the sense that actually formed distance. Vision was responsible for placing things at a distance and fixing them through perspective in their place. The visual was a means of organizing the world into objects that occupied a finite volume of space. Every glance for Welsch, had 'something of the look of Medusa: It causes objects to solidify, petrifies them. Hearing though did something completely different. It did not "reduce the world to distance," but accommodated it. If vision was a distancing sense, hearing was one of alliance' (Welsch, 1997: 158).

To demarcate vision and hearing into separate categories in this way creates a problematic binary that seeks to enact a shift in the balance of power among the two senses. For Witmore, Welsch's turn to the auditive was a symptom 'of an underlying illness that lies deeper than any supposed ocularcentrism'. Again referencing Bruno Latour, he says that it is not a question of which sense dominates, because either way human consciousness continues to be central to any engagement with the world, but rather one of human–non-human interactions in complex environments (Latour 1999). Such complex interactions are understood here as noise.

Welsch had largely ignored the important role of acoustics in the history of modernism, a history largely separated off into the specific sensory media realms of the telephone, gramophone and radio (Sterne 2003; see also Kittler 1990, 1999). Such oversight has been countered to some degree by Sterne's book *The Audible Past*. But even this important work succeeds only in rebalancing what continues to be an anthropocentric binary approach. It is precisely for this reason that the figure of noise is employed here. A temporal unfolding that moves through noise *as* noise and does not rely on the privileging of human perception and takes seriously the multi-mediating role of technology, is essential to any understanding of what it means to be human (or post-human) in the digital age. Through noise, the senses are unified and placed in close relationship to the instruments that affect and are effected by them. Noise as opposed to sound allows for a multisensual and technologically mediated temporality to be argued for. It does not segment certain specific media, as technologies of the senses, and it does not privilege human perception. Rather they are merged in a kind of phasing of time to produce a continuous discontinuity.

Jonathan Cohen also cautions against treating hearing as a unique and distinctive sense. To do so, he says, has very particular philosophical implications for how secondary qualities (colours, odours and sounds, as commonly cited) generally are understood, and in turn for the kind of engagement with the world that it is possible to have. In relation to this he argues:

> Therefore, the conclusion that sounds are distinctively temporal would be a serious blow to hopes for a theoretically unified treatment of the sensory qualities. For all these reasons, quite a lot seems to hang on the question of the temporality of sounds. (Cohen 2010: 1)

Cohen is concerned to pursue a democratic philosophy of the senses. As such he sets out to question the assertion that sounds are distinctively and essentially temporal, and in doing so also investigates the related claim that they should, unlike other sense data, be understood less as secondary qualities and more as concrete individuals or events. Commenting on the trend towards assigning such concrete status to sounds, he says, 'This has been thought to lead to the conclusion that sounds are temporal in a way that distinguishes them from non-sound sensible qualities such as color because it is generally thought that concrete individuals can, while abstracta (such as properties, and, in particular, color properties) cannot, have temporal locations and extents' (Cohen 2010: 2).

Cohen challenges the argument made on the basis of survivalist and non-survivalist characteristics – the idea that primary things and secondary qualities are variously made of either individual singular wholes (continuous) or plural streams made up of smaller mathematically individuated parts (discontinuous). The notion of survivalism refers specifically to the claim that certain things can withstand and survive significant qualitative changes to their nature and still be recognized as being the same thing. Cohen refutes the idea that survivalist characteristics inhere in sound in ways that they do not in other qualities, such as colour. The question then of whether sound is of a primary or secondary nature is an important one. The effect of assigning it temporality and individuality means that it is afforded the status of object, or as being able to be conceived of objectively. This would mean that engagement with the world through sound would produce objective knowledge, while other sensory engagement would remain subjective.

Cohen's refutation of this idea is prefaced on the belief that justifications given for the unique temporality of sound do not succeed in affirming sound's uniqueness in relation to other senses. Colour and smell may equally have temporal aspects he says. And this supports the claims being made here. If noise sound and listening are privileged here in any way, it is to address their historical neglect and lead the way for other senses to follow. If noise sound and the audible have been historically undermined because their temporal nature means that they are momentary and fleeting, then it is for those very same reasons that they are rearticulated here. It is their ability to repeat across vast timescales and to fold into other sense data that makes them so important.

The challenge then is to demonstrate how such interactions that exist between particle and particle, form constantly shifting patterns that are universal features of noise as 'simultaneous entanglement and percolation'. If this is asserted then noise can serve as an analogical device for understanding the contemporary age, and can counter the stasis that is induced by an over-reliance on the visual. The point is not to introduce noise and sound *into* critical theory, for as many have argued it has always been there and always been significant (Sterne), but to develop a non-dialectical reinterpretation of Jaques Attali's clarion call *to* noise – one that expands its definition and shows how all of the senses combine to form sonic economies wherein

> Fetishized as a commodity, music is illustrative of the evolution of our entire society: deritualize a social form, repress an activity of the body, specialize its practice, sell it as a spectacle, generalize its consumption, then see to it that it is stockpiled until it loses its meaning. (Attali 1977: 5)

From noisy beginnings then, 'our entire society' forms discernible compressed patterns that are best understood as music. Within such a society the sonic leads the way, preceding enduring, repeating and surviving over time until once discarded elements, that are always present in noise, are reenergized as the ratio of resistance in the system alternates. Music, as 'the art that it made out of time' and as an ordering of noise, in turn offers an important means of accounting for many of the uncertain and complex temporal features of our society in its contemporary guise.

The preceding pages have made reference in some detail to Paul Virilio Wilfrid Sellars, Mearleau-Ponty and Christpher Witmore in particular, as a means of beginning the process of critically evaluating the claim made by Jaques Attali that the best way to understand the world is to approach it through music. It is worth tying some of these strands together now to gauge the progress of this ambition, before proceeding directly to an analysis of Attali's proposition.

Virilio provided an important shift in emphasis from the spatial to the temporal, but never fully developed his stereophonic model. Instead, he consistently returns to an optical gaze to account for the dual existence of the real and the virtual. Sellars also set out a dual approach to knowing the world – one visible, the other invisible (mathematical and theoretical) – both very real. His aim was to merge these two poles, to tie them together so tightly that they became one. To do so was to go some way, it is claimed here, to offering a counterpoint to the concerns relating to correlationalism made more latterly by the speculative realists. Merleau-Ponty also addressed the issue of dualism through his attempts to resolve the lingering conflict between mind and body. For him the body as a receptive object is given primacy. Merleau-Ponty's body is not distinct from mind but is a fundamental organ of mind through which information travels. Yet in his efforts to reaffirm the objective body, he also reinstates the privileging of the eye, and of vision. For some it is an easy step to rethink and rearticulate Merleau-Ponty's receptive body as an immersed object that can be understood via recourse to sound, as Frances Dyson has argued. But too often this serves only to shift the emphasis from one sense to another.

The aim of what follows then is to escape the grip of the audio–visual, the mind–body, the digital–analogue and ultimately the real–virtual dualisms. Christopher Witmore makes a useful contribution to this aim. His thoughts on temporality and sound sit comfortably here. Time is described by him as a folded non-linear arena that is spatially dimensional, where patterns emerge without recourse to cause and effect and are therefore free from dialectical logic. Witmore is keen to avoid the recurring dualism of the audio–visual that shifts along the axis between the scopic and the phonic. Instead he proposes that we learn to see and to listen simultaneously. This call to simultaneity resonates here with the idea that the figure of noise can facilitate such a need – that it can account for the contemporary condition where digital and analogue forms are exchanged seamlessly, where they flow freely between real and virtual environments and spaces, folded in non-linear temporal dimensions, and where even in a time period where images seem to be increasingly pervasive and ubiquitous, they are so in a way that can be described best by using the language, representational and non representational, of noise and the sonic. Noise exists a priori and a posteriori, occurring before language and in language. It is the chaotic environment out of which language emerges and to which it returns as the entangled virtuality that is neither object nor subject, reason or experience. It is a system of engagement that operates not as bifurcated correlationism as it was for Kant, but as a social project as it was for Sellars who suggested that the ability to plot the emergence of language served to provide all the answers needed to breach the gap between the manifest and the scientific image, between rationalism and empiricism. The figure of noise can help with this as language emerges as sound and as voice. As a temporal non-linear phenomenon noise is portentous (Attali), and as word and music it moves towards meaning, although as Kant made clear, along with laughter, it did not have to mean anything.

Notes

1 This point relates to my previous work in *Chaos Media* that dealt specifically with such spatial questions and which pointed not to 'the end of space' but to the need to reconceptualize space in a digital context using sound rather than vision as an organizing principle and analogical device.

2 See Ihde (2007).

3 https://www.marxists.org/archive/thalheimer/works/diamat/10.htm.

4 There is a clear link here to the ideas of Graham Harman and his proposal that all objects ultimately turn in on themselves in silence and away from all others to form a universe of objects that are only ever partially in relation to one another. It is a universe where thought and the human mind are separated from 'the world as a whole' and where any amount of thinking can never fully penetrate its secrets.

5 https://www.marxists.org/archive/thalheimer/works/diamat/10.htm.

6 Ibid.

7 Ibid.

8 Ibid.

9 Ibid.

10 See 'The Self-Organization of Speech Sounds Pierre-Yves Oudeyer', http://arxiv.org/pdf/cs/0502086.pdf).

11 This emphasis is placed by Steven Shaviro in his *Without Criteria* (2012), and continues in his *Universe of Things* (2014).

12 See Harman (2010).

13 http://www.iop.org/news/13/mar/page_59670.html.

14 This is important in terms of what Marie Thompson describes when she warns us of the danger of expanding the definition of noise to include everything. For if noise is everything then it is also nothing. She says, 'Noise's conceptual noisiness means that it often functions as a floating signifier: it can be used to talk about almost anything' (2017: 2).

15 See, McLuhan and McLuhan (1988).

16 This comparison is explored by Carl Sachs in https://www.academia.edu/2351431/Phenomenology_and_the_Myth_of_the_Given_Sellars_Merleau-Ponty_and_Some_Myths_About_the_Given http://www.tandfonline.com/doi/abs/10.1080/09672559.2013.861002.

2

Noise and Political Economy

In his book *Noise: The Political Economy of Music* (1977), Jaques Attali presents a compelling case to support his belief that noise, sound and music should be given greater prominence as significant factors in relation to political and economic events. As an argument it is convincing, yet methodologically it relies on an adherence to dominant orthodoxies of dialectical thinking and the assumption that time is linear. This chapter and those that follow try to augment this argument by drawing on the philosophical ideas explored up to this point, and will remix these new combinations in a non-dialectical, and non-representational contemporary context that reflects on our current digital present while also projecting into the future.

Music for Attali is the organization of noise, and sound is the human perception *of* noise. Noise is the sonic chaos that unsettles and when in the process of being resolved foretells of times to come. To fully understand its significance, he set out four phases of historical or temporal development where noise operated as the ground from which music emerged as figure. His phases were: *sacrificing*, *representing*, *repeating* and *composing*.

Sacrificing

To illustrate the first phase, Attali drew on a painting by Pieter Brueghel entitled *Carnival's Quarrel with Lent*. The painting depicts a moment of conflict that conceals what might be to come: 'For concealed behind the enactment of the conflict between religious order and its transgression in festival lies every conceivable order' (21). Out of noise comes music and it emerges through a dialectical stand-off. The painting depicts relations of order and disorder, harmony and dissonance. As a painting, it is alive with noise and sound, and like in the work of Christopher Witmore, who himself used a Brueghel painting as a reference point, it speaks to a sonic environment that is both audible *and* visual.

In setting out his category of *sacrifice*, Attali referred to Renee Girard's use of that term to describe the displacement of a more general kind of violence by a ritualized form. Such a ritualized form served to tame the fear of identity that most ancient societies experienced – that is to say fear of themselves or fear of the self, as not yet fully separated from exterior things. It was a fear that led to imitation of nature and desire, rather than full separation from it, and rivalry of a kind that created uncontrollable violence. This relates to the ideas discussed earlier in relation to Wilfrid Sellars and the development of the self as a process of critical distancing that unfolded over time. For Sellars this painful process of separation would culminate in the birth of man. For Girard the solution to this uncontrollable violence (as noise), as a phase of the not-yet-complete separation, came through the identification of a scapegoat that could be ritually sacrificed. Eventually music came to be the simulacrum of that sacrifice. Noise then, in this case, stood for the murderous savagery of ancient societies. It was a disruptive state that called out to be resolved, that rang alarm bells that were so ear piercingly loud that they demanded to be softened, and that resolution, that softening, came in the form of early music. This can be read both metaphorically and literally: For violence read noise, for ritual sacrifice read music, but the state of violent chaos was also *actually* noisy, and ancient existence was alive with competing sounds and voices that have over time been calmed, resurfacing at certain moments to demand further resolution.

Noise as a category does not exist in its own right according to Attali. If noise is defined as a process or phenomenon that disrupts or undermines a more coherent or harmonious set of relations, then on its own it is meaningless and cannot exist. Noise is not a thing; not a network of things; not material in any way for Attali. It exists only as the disordered correlation of other things. A different definition of noise however can bring it firmly into existence as vibrating wave–particle event that does not rely on human perception (and as such it differs from 'sound') for its being. Noise is both a thing in itself and an interaction between things, both a primary and a secondary quality.[1]

For Attali, 'Noise is the term for a signal that interferes with the reception of a message by a receiver. Long before it was given this theoretical expression, noise had always been experienced as destruction, disorder, dirt, pollution, an aggression against code-structuring messages' (27).

Destruction, disorder, dirt, pollution and aggression: Noise has long been thought of in this way. To reduce noise to such negative concepts, practices or phenomena is however to miss an opportunity to understand how noise can operate as a positive energy force, where anomalies in a system, rather than presenting themselves as something to be removed, can be cited as sources of creative potential. Such anomalies need not be regarded as dialectically oppositional or antagonistic but as examples of difference in a chaotic and dynamic vibrating environment, where patterns form and reform, not as inevitable and teleological but as complex and not always discernible. Too often, 'as is commonly the case with things we find hopelessly confusing, noise is defined as unpleasant' (http://nautil.us/issue/38/noise/how-noise-makes-music).

Unpleasant and dirty or not, the claim here is that it does exist. It exists in its own right and as a chaotic set of relationships where, rather than being meaningless, it is precisely the site of meaning. Such a statement might serve to underpin a metaphorical or cultural use for noise as a way of bringing dark, shadowy, chaotic elements to the fore or to underpin an acoustic turn, or even to accentuate the importance of music as a significant critical indicator of social and political forces, but on a more fundamental level noise can be shown to exist on a physical quantum level. I will return to this point later.

The idea of the simulacrum of sacrifice that Attali refers to implies a deferral or a displacement. It suggests that mankind's essential and originary characteristic was based in noise, chaos and violence – states that did not immediately disappear but that resolved to find new and different forms. This idea has important temporal implications. It assumes an evolutionary linearity, a uniform unfolding as progression and replacement, rather than a singular state that finds many changing forms or patterns that exist simultaneously – patterns of recurring chaos and order where the incessant patterning is noise itself. Within such a framework, music as simulacra came to replace noise, and as representation of greater forces was characterized by the setting up of problems to be solved, of dissonances that must be resolved through harmony. To illustrate this Attali turned to Leibniz and his *On the Radical Origination of Things* saying: 'Great composers very often mix dissonance with harmonious chords to stimulate the hearer and to sting him, as it were, so that he becomes concerned about the outcome and is all the more pleased when everything is restored to order' (27).

This return to order was for Leibniz an essential feature of music as it was for Attali.

But what if it was not a rationalized representation, but a more fundamental and dynamic pattern, a code within a code?

For Attali music appeared in myth as an affirmation that society was possible. For him music simulated the social order, and its dissonances expressed marginalities. The 'code of music' served as a model for accepted rules of society. 'It is in this connection that the debate on the existence of a natural musical code and an objective, scientific, universal harmony takes on importance . . . If such a code did in fact exist, then it would be possible to deduce the existence of a natural order in politics and a general equilibrium in the economy' (29).

The search for a universal harmony was what had led Hegel and Adorno to try and perfect an ideal dialectical system. It may now be possible to locate such a system in the realm of noise as a natural code that can be artificially replicated and modelled. It need not necessarily be harmonious however, but a process that is always resolving yet never resolved. There have been many attempts to engage with nature's code as the music of existence, as the ordering of natural sounds (e.g. music concrete), and more recently numerous attempts have been made to turn the multitude of code generated within a range of different environments into digital music. From this perspective music, as the harmonization of dissonance, simulates the more basic phenomena of generalized order and marginalities, or activity at the edge of order that can potentially pose a threat or be described via recourse to entropy. Noise for Attali was the sound of things falling apart, or about to fall apart. As such it relied on these things to sustain it and bring it into being. For the purposes of this book however, noise is defined rather as the incessantly vibrating nature of all things – independent of perception. It challenges the idea of the unnecessary, the confusing, the annoying and goes so far as to say that it is precisely these 'interferences', from which occasional calm and order emerge, that are essential to an understanding of basic forces of social, political and economic life. These forces do not work in accordance with dialectical logic but as multifaceted and multidimensional features of a dynamic system. This serves as a reversal of Attali's thinking: It is not noise that does not exist but harmony (or if it does it is only fleeting, as evidenced in movement and occasional pausing, as Bachelard pointed out).

Representing

The category that Attali called *Representing* worked as a means of making people believe in a consensual representation of the world. It was a representation from which much of the 'noise' had been removed. For Attali this accounted for the history of tonal music as much as it does for the history of political economy.

'The history of music and the relations of the musician to money in Europe since the eighteenth century says much more about this strategy than political economy, and it says it earlier' (46).

Implicit within this statement was the idea that music and political economy were essentially different and separate. It describes a situation wherein the relationship of the musician to money does not constitute in itself a political economy, but is rather a sign of what is to come. There is an assumption in Attali's text that there exists a kind of pure music (at least potentially) that is not tainted by politics or economics that in turn came to be, and will continue to be, crafted in music's flawed *image*. The question is then whether music *becomes* political economy, or simply predates it? For Attali it is the latter, and he retains faith in its power as an independent realm that might still herald new forms of social organization. Music, he believes, is both a template for, and victim of, the move from traditional society to one characterized by the emergence of class based on economic exchange. So, music as a category that might be recognizable to us now, developed for Attali as part of this shift out of ritual to a simulacrum of order. The question is then, can music ever return to or rediscover its ritualistic, or pure function, or at least retain elements of it? Or is it now destined, programmed through representation, to herald only specific ends? And does that mean that it is meaningless to refer to music as a whole and further point to the requirement that we identify different kinds of music, or musics, as either tied to representation or pure and heraldic? Music, as it is understood here conforms to neither of these models, but exists as both part of, and a means of making sense of, a larger set of statements within a complex discourse that is non-linear and non-representational. It is simultaneously the sound of time and history, and its inherent instability.

Towards the end of the eighteenth century, 'music was already in contact with a new reality . . . .' (49) It was a reality that other forms of activity, perception or art, had not yet attained. Hence, it was futuristic in its orientation, a herald of things to come for Attali. But this is problematic in terms of temporal progression and the assignment of different times to different activities. It implies that music existed in one time and political economy and other forms of intellectual and artistic expression in another. But if music arrived as an aspect *of* political economy, then surely political economy itself must have also arrived. History thought of in this way becomes staggered and phased, as the musician, the philosopher, artist and the bourgeois consumer are positioned in linear counterpoint until time eventually catches up with itself to the degree that a new dawn or age is realized. This phased nature of temporal transition is described by Attali with reference to Mozart, who was appalled by the status of music during this period of uncertainty when it was no longer a privileged aspect of aristocratic rule, but shackled by a new kind of ownership. For the newly emerging bourgeoisie, music was a possession, something to have rather than something to listen to. So, the resolution of noise as organized music enacted and made distinguishable different measures of time that may or may not have been relative in terms of tempo as one kind of activity eventually syncs with the others. Such relativity implies a phased continuity, whereas isolating the different measures implies a discontinuity where different histories can develop at their own pace, each unaffected by the other. Such a contradictory position requires a theoretical framework that can account for music not as an abstract independent activity with the power to foretell, but as one among many interrelated activities that feed into and out of each other as complex orchestrations.

Music during this period had escaped the aristocratic grip of feudalism and begun to develop as an economic commodity that could be owned and traded on a systematic basis. This situation prevailed during the early modern period and up to the French revolution when a brief attempt to protect the interests of the composer and harness the revolutionary power of music was initiated (55). In 1793, the production of music in France came under state control with the intention of 'annihilating the shameful torpor into which [the arts] had been plunged by the impotent and sacrilegious battle of despotism against liberty' (55). Music it was believed could both protect and fortify the revolutionary zeal at the outset of the modern period. Music would function in the service of state ideology oriented towards the betterment of society, performed on a grand scale in public, available to all and released from the yoke of bourgeois propriety. It was to be a soundtrack to the future a *future sound*. But the utopian strategy never fully matured, and music quickly fell under the control of the newly rich, and continued its journey as valorized commodity. The commodity took the form of paying to listen.

Musicians no longer sold their bodies in terms of committing themselves to a single patron, but instead received remuneration in return for their labour, as and when it was enacted in the service of whomever was willing to pay. This brought a new-found freedom for musicians and composers, and music offered up a model of the future: 'Thus delimited, music became the locus of the theatrical representation of a world order, an affirmation of the possibility of harmony in exchange. It was a model of society, both in the sense of a copy trying to represent the original, and a utopian representation of perfection' (57). It was a model of potential, music was showing the way, pointing towards an idealized future, in this case characterized by bourgeois individualism. It is the idea of music as a model of potentiality that is significant here. And if Attali's logic is to be followed music set the scene for a quarter millennia of capitalist domination. But what, if anything, do contemporary changes signal? Is there a return to representation as new economic models develop in relation to music? These questions will be engaged with more directly as we proceed.

Prior to the period of representation, music had been inscribed as the 'background noise to everyday life'. 'When people started paying to hear music, when the musician was enrolled in the division of labour, it was bourgeois individualism that was being enacted: it appeared in music even before it began to regulate political economy' (57). Again, Attali cites music as a very broad category that prefigures all other forms of activity. But, as he himself points out, this form of publishing, now applied to music, had appeared in print even earlier. Music publishing came out of bookmaking, and as such was implicated in a far more complex and nuanced set of relations than Attali is able to account for (52).

Music had entered the realm of exchange in the form of theatrical performance. Attali argued that this exchange was inseparable from the performance as part of a process of 'making people believe' in order, or in the possibility of a rational ordering where ' . . . the utility of music is not to create order, but to make people believe in its existence and universal value, in its impossibility outside of exchange' (57). Outside of exchange, outside of the economy, was where chaos reigned. The implication of this is that order is impossible outside of a system based on exchange where order and economy become synonymous. Music and its development in the eighteenth century, was interpreted as having promoted and supported this idea. It was a process that involved the notion of exchange and a logic that demanded reliable values and equivalences. The process had to be rational and reliable, with the result appropriate to the promise (a process similar to the linguistic system as defined earlier by Wilfrid Sellars, where propositions had to be substantiated). As such the economic activity of exchange was understood as a *bringing to order* and it had the related effect of marginalizing disorder. A sonic economy on the other hand – of the kind that is being argued for here – undermines the order of certainty and reintroduces uncertainty as a common feature. It brings back that which has been excluded.

The economic realm of exchange that was heralded by music assumes a notion of value in things prior to their exchange. What is represented, Attali says, 'must be experienced as having an exchangeable and autonomous value, external to the representation of the work'. The thing, the work, in and of itself must be recognized as having value. It can then be represented and exchanged accordingly. But how is this kind of value measured. The Marxist response has been to draw on the common denominator, labour. In the musical realm, the 'work' is the composition, the manuscript or score, and the representation is the performance. 'Music, [previously] meaningless outside of religion, takes root in representation and therefore in an exchange of labour allowing a comparison between representations to be made' (58). But what if the 'work' is itself a representation? The thing represented in turn must also be assumed to have a value, even if it is of a different order and required no labour, such as nature or emotions for example. This kind of value cannot be measured nor exchanged and thus lies outside of formal political economy for Attali, but it is not outside of a sonic economy that includes musical expression among a range of other intangibles in it accounting.

In short, for Attali, music entered the realm of money – itself an abstract realm of generalized and predictable exchange.

> Representation entails the idea of a model, an abstraction, one element representing all the others. It thus relates to the political and the imaginary, but first of all to money, the abstract representation of real wealth and the necessary condition for exchange. The idea, which was very new at the time, that it is possible to represent a reality by a form, a semantics by a syntax, opened the way for scientific abstraction, for the attainment of knowledge through mathematical models. (58)

Music preceded claims relating to the algorithmic nature of all reality, and the eventual sophisticated development of models capable of accounting not only for mathematical representations *of* reality in the abstract but for the actual mathematical nature of the real as an 'economic realm'of intraction and exchange'. But if music paved the way for mathematical models of reality, so mathematical models would return to inform music. As an example of relative non-linear phased temporality, the algorithm, as will be discussed later, would become an essential feature of music in a range of technologized forms, as it would for the economic realm and eventually our 'entire society'.

For Attali, the circumstances of man find their expression in music before they find their equivalent form at the level of political economy, and there is a direct correlation between the two. But again, they are understood as separate. This is a linear relationship that has a specific temporal dimension. The problem arises when thought begins to operate through music and not just in relation to it. This is what Attali set out to do, but he was not always successful. Music seems to fluctuate between being a category in its own right and an aspect *of* the wider political economy that it heralded. If music adopts features of what will become political economy is it not therefore preceded by a political economy up to that point unnamed – a consequence rather than a subsequence? The model for what will become political economy infects music, which in turn begets political economy itself. This problem is at the heart of a much wider set of questions then that relate to temporality, succession and dialectics. This is why a temporality of noise is proposed here. It overrides problems of linear succession without recourse to cause and effect.

Thinking through music also throws up the problem of which music we should think through? Music comes in many forms; many different patterns form around it as a single nomenclature, and each has a unique and different relationship to its circumstances. For Attali such decentred and eclectic practices lie outside dominant forms of music and therefore constitute a sacrificial form of music as noise. He may be right, but nonetheless this is where this investigation is situated, and to make any sense of this multifaceted noisy realm, political economy must be replaced by a sonic economy that accounts for the unaccountable. The music that is 'meaningless outside of religion' may turn out to be a very rich source of meaning after all.

Through representation music achieved an exchangeable value. It moved into a phase where it could be coded, and could be both quantified, qualified and copied, to a degree. But once again the question arises that could music, or a form of music, continue to operate outside of economic value – impossible to quantify, the stuff of the soul, as affect, and beyond representation? Of course, Attali was aware that music had a 'use value' of a kind that did not suit quantifiable measurement or its transformation into political economy. 'Nevertheless, representation was able to make people believe, for two centuries, that it was meaningful to have a measure for value, that exchange and usage existed and came together in value' (59).

If the wild complexity of noise could be tamed as music, and the intense affective nature of music could be tamed as representation of order, then all aspects of life must surely follow. This taming took its form in harmony as the avoidance of violence and noise. Such ordering 'presupposes a topology and mathematical model: the mathematics available at that time was necessarily based on theories of the machine in equilibrium, in harmony. Here again, music prefigured the trap into which the major part of political economy was to fall, and where it would remain to the present day' (59).

In assessing the mathematical nature of his representational model, Attali made the distinction between nature and science in terms of the assignment of music as a means of imposing order as harmony. The harmonic scale, like the machine in equilibrium, he said, was the incarnation of the harmony between heaven and earth, between the absoluteness of the Gods and the simulacrum of earthly representations (another duality). This referred to the use of science to resolve the uncomfortable aspects of nature – to get everything under control and enrolled in a system of sense making. Harmony, thus, became an ideology in which dissonance had no part other than to reinforce harmony as the better option. This negative situating of dissonance, that is echoed in Adorno's work, is reconstituted here through the figure of noise as being indicative, not of an outside irritant to be placed at a distance, but of a virtual ground from which everything emerges – not as effect but as the dynamic becoming of the manifest through the scientific.

The application of science and mathematics to music, Attali said, would go on to find its ultimate manifestation in the field of political economy, and would do so in an ordered and predictable manner. Attali was sure that a degree of predictability would have been possible in the eighteenth century, if only someone had known what to look for.

> By observing music at the end of the eighteenth century, or at latest toward 1750, one could have made a serious prediction about the subsequent evolution of the system and about its limits. To make people believe in order through representation, to enact the social pyramid while masking the alienation it signifies, only retaining only its necessity – such was the entire project of the political economy of the last two centuries. (61)

The use of the term 'observing music' is interesting here and once more undermines Attali's stated aim to think *through* music. Music was cited by him as a model *of* political economy, as a representational code that could be applied in turn *to* political economy. This implies that music for him occupied a position in between one form of political economy and another, operating almost as a bridge or dialectical enabler. If music was an interim phase then it would always have been possible to ascertain what would come next, so long as one could sufficiently decode the available data.

'Before political economy, then, music became the bourgeoisie's substitute for religion, the incarnation of an idealized humanity, the image of a harmonious, nonconflictual, abstract time that progresses and runs its course, a history that is predictable and controllable' (62).[2]

If music instantiates this degree of predictability and control, what can usefully be gleaned from it? What might the current state of music tell us about what is to come, or even more significantly maybe, about the time we are in? For the question must also be raised as to whether music is only significant at times of imminent change? If music precedes, how are we to know what music heralds and what music does not, without knowing which events are significant and which are not? In short, must the exercise always be retrospective, as is the case with Attali? Presumably if music stays the same then nothing will change. But by what criteria should we recognize a shift great enough to pre-empt change, and how do we translate (decode) the nature of that shift so as to be able predict the change to come? Once more the digital paradox may be significant. If currently our perception of, and immersion in, time constitutes a peculiarly digital, non-linear, multi-temporality, then perhaps it can only be ordered, patterned and 'understood' using digital means. We now possess the kind of mathematical and scientific models necessary to enact such an ordering. Techniques based on fuzzy logic and neural networks can now be used to map musical characteristics and their relationship to social and political events in increasingly complex ways. The possibility may exist therefore, to test Attali's propositions through mathematical means, and although doing so is outside the scope of this book, the long-term plan is to do just that.

Through such digital means, we are now capable of measuring and documenting changes to music itself, and not just the structural changes in what might be called the 'music industry', or the 'political economy *of* music'. It is possible to measure with great accuracy modulations in the dominant harmonic model. It may even be possible to use such modulations to predict the end of capitalism. Although this would seem to be a very crude measure, and constitute a vastly oversimplified reading of Attali's predictive model. But it might be possible to discern some interesting information from a more nuanced model that matches discrete shifts in pitch, rhythm, timbre with events that might be said to come under the heading political economy. Attali himself does not do this. As Josh Epstein says in his book *Sublime Noise*: 'If Adorno is proudly narrow minded, lovingly patient with the music he admires and magisterially dismissive of the music he doesn't, Attali proves trickily unselective, neglecting – for all his rhetoric of liberated composition – the interpretive operations of the specific artwork' (Epstein 2014).

In other words, he never really succeeds in thinking *through* music, but only in ascertaining where it stands in relation to political economy. To satisfactorily think through music and not simply about it would necessitate an engagement with what Attali himself refers to as *combinatorics*. He says,

> As in music, combinatorics in production is thus central to the search for and formation of compromise, of harmony between divergent interests. But combinatorics is only possible in the limited field of discrete and controllable sounds. Beyond that, it gives way to statistics, macroeconomics, and probability: before the consumer did, music demonstrated that combinatory growth explodes in the aleatory and the statistical. (65)[3]

It was evident, according to Attali that the combining of discrete elements into a discernible pattern or system occurred in music before it took hold in the area of political economy. It signalled a bringing to order within a limited, or closed system. All that lay outside of the system was a threat to it (that which had been discarded, as noise according to a conventional understanding of what is meant by noise) and once music, and by association political economy, reached the limits of the system, the drive towards new forms of expression would increasingly step outside of that system – stepping back into noise in search of new forms of knowledge and creative expression. However, this idea of an inside and outside of a system may be limiting as it is always deferring to a dialectical rationalism and logical succession. A better way of understanding this relationship is to regard the inside and outside as constant aspects of one another, as noise and entropy.

When music reaches the limits of its capabilities within a given system, Attali suggests that new solutions are sought, and there exists a thoroughly documented history of experimental interventions, refusals, proposals, that plot the trajectory of 'new' forms of expression in music throughout all of Attali's named phases. But these largely represent a statistical and aleatory diversion away from music in his opinion. It is just such statistical and aleatory diversions that are of interest here. They seem on reflection, however, to have signalled little if anything in relation to the wider political economy since its inception. But that may be because by stepping outside of music, as an established system, these experimental forms had to forgo the heraldic ability that is assigned to music. A more likely explanation is that these experimental forms accompanied rather than heralded a morphing of political economy: A transformation that has come in the form of an increasingly complex and expanded combinatoric field with its continuous discontinuities forming, deforming or reforming as information in a dynamic cosmos that is increasingly able to be modelled while at the same time challenging the limits of modelling and representation itself.

Whether or not radical experimentalism steps outside of an organized system that can be called music to such an extent as to undermine its status *as* music is a key point of conjecture. For Attali such a radical step was ultimately meaningless and offered no herald. In contemporary terms, however, the very idea of an expanded combinatoric field may itself have been pushed to its limits in an era where infinite combinations are possible at ever-increasing speeds, creating noise of uncontrollable and unpredictable proportions on a permanent basis. This would explain the always approaching, always fading, imminent catastrophe or permanent catastrophe that sometimes seems to be a characteristic of the digital age, as it had been a century earlier. In *The Rest is Noise* (2007), Alex Ross describes Europe at the end of the nineteenth century as being characterized by just such a sense of 'imminent catastrophe' (40).

The end of the nineteenth century was a time when occult and mystical societies were prevalent as people tried to make sense of the 'changing times'. And it was in Vienna where the changing times seemed most evident.

> Vienna was the scene of what may have been the ultimate pitched battle between the bourgeoisie and the avant-garde. A minority of 'truth-seekers,' as the historian Carl Schorske calls them, or 'critical-modernists,' in the parlance of philosopher Allan Janik, grew incensed by the city's rampant aestheticism, its habit of covering all available surfaces in gold leaf. They saw before them a supposedly modern, liberal tolerant society was failing to deliver on its promise, that was consigning large parts of its citizenry to poverty and misery. They spoke up for the outcasts and the scapegoats, the homosexuals and the prostitutes. Many of the 'truth-seekers' were Jewish and they were beginning to comprehend that Jews could never assimilate themselves into anti-Semitic society, no matter how great their devotion to German culture. In the face of the gigantic lie of the cult of beauty – so the rhetoric went – art had to become negative, critical. It had to differentiate itself from the pluralism of bourgeois culture, which, as Salome demonstrated, had acquired its own avant-garde division. (Ross 2007: 40)

An approach committed to noise similarly speaks up for outcasts, and it does so at a time not unlike the one described earlier when promises seem to have been broken, and when popular response seems to be manifesting itself in a sharp shift to the right of the political spectrum. But it is not truth that is sought, but a reset, remix or reorientation of approach, a recombination, re-patterning of the elements that assemble and disassemble in such a way that no promises make sense, or indeed in a way where promises are understood as the *making of sense* through the contextualization of a future that is believed in but never delivered.

In late nineteenth and early twentieth century Vienna, it was architect Adolf Loos, artists Oskar Kokoschka and Egon Schiele and poet Georg Trakl (Ross 2007: 41) who were pushing artistic forms in new directions. But ultimately Ross, like Attali before him, concludes that the avant-garde promise of modernism was a surface phenomenon, a *representation of critique* that quickly morphed into something far less revolutionary. The entire discourse surrounding the Viennese avant-garde, he says demands skeptical scrutiny (42). This was particularly the case with reference to the work of Otto Weininger whose work was prevalent and influential during this period. Having commited suicide in 1903, Weininger's racist and misogynistic ideas were of considerable influence during this period. 'As in prior periods of social and cultural upheaval, revolutionary gestures betray a reactionary mind-set' (42).

The idea of a bourgeois avant-garde resonates strongly here. Is there a contemporary parallel? Who might fit into the category of bourgeois experimentalism? The temptation to make a list is probably best avoided. Tempting as it is, the intention of this book is not to assign artists to an in–out process of assigning authenticity, but to take the entire sonic environment as a noisy and unpredictable realm where patterns are discernible; patterns that might speak to us if we are able to tune in and decipher.

In music generally Ross says, things were a little different: 'If the ethical justification of the modernist crusade rings false, composers did have one good reason to rebel against bourgeois taste: the prevailing cult of the past threatened their livelihood' (2007: 42). There are a number of important points to be made here: Music is regarded as having retained a level of authenticity when all around it was a suspect, and it did so from within a framework of political economy, and also within a temporal framework. It was aiming forward while being pulled backwards. In Vienna of the early 1900s music had settled into a canonical repertoire of classics and populist new forms that led Schoenberg and Mahler to the point of despair. The status of forward thinking music, or *future sounds*, had been called into question. This struggle between artistic expression and the need for an audience was significant. 'Fin-de-siecle Vienna offers the depressing spectacle of artists and audiences washing their hands off each other, giving up on the dream of the common ground' (43). This is a repeating pattern, and there is a long history of musicians making music that has no audience, or that deliberately alienates the audience, or produce music that the audience is not yet ready for. From such a perspective, the heraldic feature of music takes two forms: The one where it veers off on a popular trajectory of repetition, and the other where it is a portent of a promise unfulfilled. This battle between the popular and the experimental would prefigure, Attali suggested, the path chosen by the wider forces of political economy for another 100 years.

By the end of the nineteenth century, the period of representation as held together by harmony was beginning to unravel. It had been a system where representation took the form of performance and wherein the music and the rights and status afforded to musicians and composers had prefigured the dominant mode of capitalism, enshrined in both law and practice. The epicentre of the unravelling was once more Vienna. Mahler, Wagner and Schoenberg were pushing the system beyond its limits. It proved however not to be a utopian unravelling, but a 'drift towards repetition', a new and more efficient way of organizing and ordering even where disorder seemed to be evident. The system would find a way to neutralize its revolutionary potential by a process of dilution where experimental attempts to recognize and celebrate chaos and complexity were eventually adopted by elites (Adorno would later reflect on this process in his *Negative Dialectics*).

'In fact, music at the end of the nineteenth century was highly predictive of the essentials of the ruptures to come. And practically everything that happened took place in Vienna . . . The present economic crisis and efflorescence of our decadence were programmed in Viennese music' (Attali 1977: 81).

Just as chaos was being celebrated and embraced by new forms of music however it was already in the process of calming itself once more:

> In the early 1900s, amid the chaos of a world at war, the psychologist, Max Wertheimer, investigated the underlying principles that enable humans to find structure in noise. Wertheimer's gestalt laws of perceptual organization described how the mind groups disparate objects if they are proximate or similar. He demonstrated how patterns can be inferred if they suggest connection or continuity. Wertheimer termed the fundamental principle of gestalt perception *prägnanz*, as the tendency to seek simplicity through recurrence, order, or symmetry. (http://nautil.us/issue/38/noise/how-noise-makes-music)

Once again music seemed to overpower noise as it moved to its next phase.

Repeating

What was started in representation was finally accomplished in repetition. Any remaining trace of the function of representation was illusory and designed to ease the transition. For Attali, the development of recording technology had a seismic, if unintended, impact in this regard.

> This constitutes, moreover, a massive deviation from the initial idea of the men who invented recording; they intended it as a surface for the preservation of representation, in other words, a protector of the preceding mode of organisation. It in fact emerged as a technology imposing a new social system, completing the deritualization of music and heralding a new network, a new economy, and a new politics – in music as in other social relations. (89)

If such technology had been intended as an advanced means of stenography, it would be its application to music that would be most significant. Representation had developed in the eighteenth century and taken hold in the nineteenth century, but as the twentieth century approached an important change in affairs was about to become apparent.

> This radical mutation was long in the making and took even longer to admit. Because our societies have an illusion that they change quickly, because the past slips away forgotten, because identity is intolerable, we still refuse to accept this most plausible hypothesis: if our societies seem unpredictable, if the future is difficult to discern it is perhaps quite simply because nothing happens, except for the artificially created pseudoevents and chance violence that accompany the emplacement of repetitive society. (90)

Such a transition can be understood by invoking the contemporaneous nature of the slow and the fast – the glacial and the rapid. Change is experienced as a radical discontinuous break, but in actuality it forms part of a continuous temporal chain. To fully appreciate this apparent paradox requires that we engage once more with Alfred North Whitehead, for whom process, movement and decay are as valid for those entities like mountains, which appear static, immovable and permanent as they are for a mayfly whose demise begins almost immediately after its being begins. Both of these states or modes of time exist simultaneously as *enduring objects* without one necessarily troubling the other (Whitehead 1929). So essentially time cannot be understood as a single coherent concept, but must take account of the commingling of different modes. But with a specific mode of time in mind, Virilio's 'world-time', it is important to consider the question of whether the world is speeding up or slowing down? In asking this question we might refer to the speed at which life is lived or to the ordering of its passing. This depends on whether we are talking about duration (lived) or mathematical (measured) time. As a means of answering this question, we might go back fifty years to 1967, and back from there to 1917 and try to measure the differences in musical-time – from psychedelia to contemporary music it seems like almost no change has occurred, but from Dixieland jazz to psychedelia it seems like a lot of ground was covered in the same measured period. So maybe, counter-intuitively, things are actually slowing down – at least at the musical level. A more useful explanation would be to embrace the Whiteheadian notion of time as a multimodal, multi-tempo and relative concept that does not conform to a universal model.

According to Attali's description *actual* and significant change goes unnoticed or is shrouded in a surface chaos of perceived *constant* change and unpredictability that occurs largely on a visual plane that is not a reliable register of genuine transformation. This is a fairly good description of the contemporary digital landscape with its constantly shifting and rapid visual manifestation where despite the appearance of change in the form of pseudo-events, things stay largely the same. Taking notice of the change is for the most part something that is done retrospectively. And this raises once more the question of whether music can only ever be assigned premonitory status a posteriori? Going back in time to validate the predictive powers of music in relation to events that have already happened might, with some justification, be treated with a degree of scepticism.

Nonetheless, let us continue. Attali takes a dim view of repetition, no doubt with some justification. But a closer analysis may raise a few interesting counterpoints. He described a capital-intensive industry where technology played a significant part as music once again ushered in a new kind of capitalism. It was an industry where production and distribution of records was an expensive business which required considerable and concentrated investment.

It heralded the transition from competitive capitalism (representation) to monopoly capitalism (repeating). The economy of the music industry was distinct and once more heraldic in so far as its products were more than just the objects that it produced – more than the records and more than the work the records mediated. The use value in the case of recorded music was a complex scenario in which objects of desire as technology were created to carry content, the desire for which had also been created by an industry that was aware of the necessity for the production of a set of conditions wherein its business could thrive.

'It is thus essentially an industry of manipulation and promotion, and repetition entails the development of service activities whose function is to produce the consumer: the essential aspect of the new political economy that this kind of consumption announces is the *production of demand*, not the *production of supply*' (Attali 1977: 103).

It was for Attali a history best told through the appropriation and commodification of a range of music, from jazz to rock where the lived, embodied and often rebellious nature of its production was shaped and patterned into a palatable commodity for potential consumers whose tastes had been similarly moulded and patterned. But it may not have been as simple as this narrative suggested. Attali recounted the appropriation of jazz by predominantly White commercial interests, and extended this account into the subsequent development of rock and pop music. These histories have been accounted for elsewhere in countless volumes, and the point here is not to contribute any further to this, but instead to ask what we might learn from listening to such music rather than simply aligning it with political and economic trends? As a model of political economy Attali's account undoubtedly has some validity, but it tells us little of the music that he described as having been appropriated, its long history and sudden shifts. It seems theoretically difficult to argue against the ability of capitalism to appropriate everything in its path, to colonize it and render it devoid of meaning. It does do this. It has no scruples. There is nothing it will not devour (including itself). Yet in the midst of this devouring, music did more than simply herald, it mutated as noise as it emerged to disquiet and unnerve.

Attali thought differently. For him, under the regime of repetition, the nature of music as a technology-driven manufactured process involving producers, engineers and listeners stripped music of some of the qualities it had demonstrated under representation: 'Little by little, the very nature of music changes: the unforeseen and the risks of representation disappear in repetition. The new aesthetic of performance excludes error, hesitation, noise . . . This vision gradually leads people to forget that music was once background noise and a form of life, hesitation and stammering. Representation communicated an energy. Repetition produces information free of noise' (106).

The patterns formed in representation had now become engrained for Attali, deep and intractable as music became *noise-free* information. Where exceptions to this dim view of music occurred, they were quickly censored and suppressed, he thought. This argument is not necessarily convincing. For music, as an aspect of noise rather than an *ordering of*, is far more nuanced in its character and affect. Music itself is not a uniform or general category. When treated as such, it tells us little of the world into which it emerges. Rather, as a complex system of fluid patterning where difference repeats, it may tell us something about the temporariness of its existence, as indicative of a form of engagement where a constant feeling of uncertainty is a positive, and where the changing of mind is a virtue. It is a form of non-dialectical engagement where systematically retaining the values and aesthetic modes of the present as one moves into the future can never be a good idea. Music that is cherished today may be recognized as incongruous tomorrow. Equally it may be revisited – dragged from the past, recent or distant (like Deleuze's festivals that repeat), to perform a new role. This is the temporal value of music beyond its existence as a mere product to be exchanged.

The result of the dialectical shift from representation to repetition for Attali was the triumph of organized capital – in music before any other realm – that was comprised of a dominant system of performers and practices that made up the recording industry. There may also have existed, he claims, a residue of local artists who performed a different function, and who retained an element of representation and festival, of music as background noise, music as lived. At times this 'apocryphal realm of the amateur' (Benjamin 1970), acted as a feeder stream for the professional sector, but more often for Attali its function was quite separate. Theirs' was a practice that occurred outside of the system and thus not indicative of future trajectories.

The relative status of professional and amateur may have been altered with the onset of new technologies in the 1980s and 1990s, as independent labels began to proliferate. Yet history shows that they too were quickly assimilated into a system of repetition. Once again there have been numerous accounts of this failure, and there is little need to repeat them here. What does need to be stressed here although is that what is required is not an analysis of music from a political economy perspective (as Attali has done quite brilliantly), but a new model of political economy itself, that uses noise, sound and music, not as indicative features of an existing model of thought but as primary features that underpin a new model.

Music, in the mode of repeating, had no inherent value beyond the constructed spectacle that is disseminated through the artificial mechanism of supply and demand, where difference, flattened as production was standardized, and consumption was based on artificially assigned value (in the case of music, chart ranking). When everything sounds the same and there is so much choice that choice itself becomes meaningless, capitalism reaches its purest form as repetition. Forty years after Attali first described this trend, it may seem to read truer than ever. Attali was describing the birth of popular music as the foreword to a new kind of advanced capitalism – one without depth or value and where the simulacrum preceded. The music of the 1950s and 1960s heralded this move. What was happening musically predicted what would be happening 15 to 20 years hence, or up to the time of Attali's writing. Whether we are still there or whether the intervening years have been witness to more predictions or predictive events is the central concern of this book. The problem with this model although seems to lie with the universal belief that manufactured music is meaningless in itself. I think this assumption is worth challenging, or at least considering further.

Attali went on to say that hits are not produced by the industrialized nature of production but are validated by the mediated system of distribution and organization they enter (108). Hits are not 'fabricated' but the very fact of being a 'hit' means having entered a managed system of value attribution that impacts on both future supply and demand. There is no way to manufacture a 'hit' he says – even 'manufactured music', and there is plenty of it, is not necessarily a 'hit'. The music can be manufactured but the 'hit' is systemic and subject to the arbitrary attribution of value. The system does not care if a product is any good or not – it cannot because inherent value within the mode of repetition has ceased to exist. There are too many releases to reliably predict what will be a 'hit' or not – so a degree of statistical chance enters the equation. No matter how much is known about the variables there is always an element of the unknown, a degree of chaos. What was key for Attali although, was the ability of advanced capitalism to absorb the unexpected, to integrate it quickly into the system.

This notion of a 'hit' that came to characterize the music industry was also present in other aspects of political economy, and was a model for the operation of advanced capitalism, and the vacuous surface existence of pseudoevents in a world where nothing happened any more. 'We can even go so far as to say that since the emergence of the hit parade, all that radio broadcasts anymore is information: on the spectacle of politics in newscasts, on objects in advertisements, and on music in the hit parades' (Attali 1977: 108). And this idea of *the world as hit parade* points to the notion that superficially at least, this kind of unpredictability is almost constitutive of a utopian socialist economy where participation in the market is equitable and fair (107). Alas this is only an illusion and the reality for Attali is that there is complex management of both supply and demand that undermines any genuine participation. Such a pattern would once more be repeated in other aspects of political economy that would follow the mediated network logic of music. But yet again music is *viewed through a len* of political economy; it is the political and economic aspects *of* music that are to the fore, and not the music itself.

Once a piece of music enters the charts it is for Attali revalued, if not devalued. Essentially it is an act of 'turning rebellion into money'.[4] Some products are designed for the system while others are appropriated by it. Others still are designed to circumvent it – though that in itself cannot ensure they are not appropriated. The system can easily accommodate the flattened uniformity of manufactured pop, and the uniqueness of experimental output (as surprise hits).[5] So then in the mode of repeating music operates as a preceding analogue of the wider economy. Young people are separated from adults, not to accentuate difference but to impose similarity.[6] Music is the realm where they are socialized, where they learn to be consumers, responding to and simultaneously supporting a demand that has been managed and ordered. It is a demand for and consumption of, a banalized supply of standardized featureless offerings. Yet there always lingers the possibility of a significant 'event'.

'At times, however, the quality improves, songs become critical and music blasphemous: *repetitive*, *detached*, as though denouncing standardisation; it heralds a new subversion by musicians cramped by censorship, who stand alone in announcing change' (109).

The notion of a qualitative improvement is problematic though. Having outlined a world modelled on the hit parade where nothing happens anymore, and where capitalism can absorb all that threatens it, we seem to be being encouraged to listen out for the possibility of change to come. And if change is to come it will once more become evident in music first. But it will be in the music of the previously maligned, and necessarily reconstituted avant-garde, and not in pop music, Attali asserts. He takes a fairly dim view of pop music describing it as the confinement of youth (109) but there is finally a sense that Attali is listening (although not giving examples). There are signs here at last that he is enacting his own belief in a new methodology where music is employed to understand political economy and not the other way around. Music, if its pure form can be found, might once again herald. Yet one suspects that evidence of change will be sought in the changes to the political economy *of* music, its mode of production, distribution and consumption, rather than in the music itself. Yet political economy is itself a sonic phenomenon where patterns form in multiple temporal dimensions and music is one of the sounds of political economy not its antecedent. It comes in many forms and is an aspect of noise rather than simply its ordering, and it is full of information, not an indication of another more meaningful kind of information, but information itself in visceral affective form. This is noise as everything all at once: The subject, the object, the scientific and the manifest, the mind and the body, the past, the present and the future, the senses. Whether in twos, threes or fives, they entwine and entangle through resistance to create a singular and dense material plenum from which emerges and submerges all life sentient and otherwise.

But despite noting the potential for new and experimental forms that emerge as music, Attali did not overall support the efforts of the theoretical vanguard. Their work, he thought, was abstracted to the point of rendering it meaningless. Rather than challenge the mode of repetition it too often reinforced it, as part of a strange alliance with mass music. For those at the forefront of experimental practice, music had to seek a renewed status. Attali called this *scientism*. Referencing Pierre Boulez and Iannis Xenakis he described how future music would strive to undermine the idea of a criterion for truth or a common reference. Music would return to noise. As music came to embrace complexity and uncertainty, so too would other aspects of life – political economy and science. Again, Attali was trying to draw temporal connections instead of recognizing that this *future music* was a sounding of the reconfiguration of patterns that had come apart, only to form or reform anew – not as dialectical progression but as non-linear non-representational vibrating matter.

Scientism, imperial universality, depersonalization, the deconcentration and manipulation of power and *elitism* were all features for Attali, of the attempt to forge a new kind of music – or a *future sound*. They were evident in music and preceded eventual characteristics of political economy once more, he thought.

In relation to *scientism* he said,

> The parallel to science is total. Like science, music has broken out of its codes. Since the abandonment of tonality, there has been no criterion for truth or common reference for those who compose and those who hear. Explicitly wishing to create a style at the same time as the individual work, music today is led to elaborate the criterion of truth at the same time as the discovery, the language (langue) at the same time as speech (parole). Like science, music then moves within an increasingly abstract field that is less and less accessible to empiricism, where meaning disappears in abstraction, where the dizzying absence of rules is permanent. (113)

'Criterion for truth' and 'common reference', although admittedly present in some forms of tonal music, are largely visual concepts and belong to the ontic ocularcentric world in terms of a bringing to order, as discussed earlier with reference to Christopher L. Witmore, and to be discussed later with reference to Sterne's audiovisual litany. So, invoking them to critique experimental sonic practice seems a little misplaced. As does the accusation that language (langue) was developed at the same time as speech (parole) – something that could just as easily be levelled at Attali's assignment of music's predictive qualities after the event, where the beginning is written only once the ending is known.

Without tonality as an ordering principle Attali thought, music had descended into noise. It was not that music had challenged accepted criteria for truth or common reference but that music, without the ability to account for itself, had become meaningless *as* noise. It had no anchor, at least not one that could up to that point be expressed. But what if finding an appropriate anchor or ordering principle was not necessary? Patterns will always find a way to emerge, and as such music as raw expression is an important but not unique example of a world that emerges from and is immersed in the vibrating energy of noise without teleological boundaries. The idea that all meaning has disappeared into abstraction, is not necessarily a frightful one. The idea that meaning is constantly forming and deforming without recourse to an accepted standard in a universe where humans are just one element in a much wider cosmology is embraced here. Where Attali sees the 'difficulties of a science of repetition' as meaningless (113) I see opportunities. As Attali was striving to identify heraldic moments, the intention here is to listen for the sound of a universe reconfiguring itself endlessly. As was the case in 1900 viewed from Vienna, the world today seems to be at the point of impending crisis. Established anchors and ordering principles seem to be less solid. The system has reached its limits, it seems, and challenges are beginning to form outside as dialectical antithesis – or so it would *appear*. An alternative hypothesis, arrived at through listening, is that the harmonic bonds have begun to weaken allowing dissonance to emerge in a complex environment where multiple and dispersed aspects of the entire universe coexist as noise. No longer can closed and exclusive systems of thought impose themselves on such complexity.

Referring to the idea of *imperial universality* Attali claimed that theoretical music sought to account for everything. It is described as an elite practice without an audience that imposes itself on a small but powerful corpus of society. Its aim was to establish music as a kind of theory of everything. Music could be anything and it could be everything, it had no specific set of codes that it needed to adhere to and no set goal beyond its drive towards universality. Such universal practice was epitomized for Attali by artists such as Xenakis and Karlheinz Stockhausen among others. Contrary to his dismissal however, this drive towards a theory of everything is regarded here as a potential strength rather than a weakness.[7] It is once more the 'everything' all at once that constitutes noise (Brassiere 2007).

For Attali theoretical music was ultimately meaningless. 'The music of power no longer conveys information within a code. It is, like the ideology of the period, without meaning' (114). And this may also be a danger for the idea of noise that is being proposed here. That is if the pursuit of meaning is the ambition. There is always a danger that when a concept attempts to account for everything, it can end up accounting for nothing. This danger is noted, but noise can operate as both a general and universal description of everything as vibrating energy and as a more specific description of that which is deemed unnecessary or unwanted. As such, and by recuperating elements that do not conform to a dominant code, it opens up closed systems of rationality and logic like dialectics to approach a very real world of complexity that can be encountered without being fully accounted for. Undoubtedly Attali would not concur. For him the theoretical musicians were in search of a project that could be understood as the futile pursuit of a new universal language that would bring meaning. He described this project as: 'A frenzied search for universal abstraction by men whose labour has lost its meaning and who are incapable of finding a more exalting one for it than statistical organisation of repetition' (114). Mapping and recording the 'statistical organisation of repetition' may be exalting reason enough. Ultimately, this meaninglessness in so far as it is not closed off to future possibilities, does contain within it the notion of potential. Free from the constraint of a dominant code (configuration) it is able to turn into something positive as difference (Deleuze 2008). What that is or was remains to be determined (or maybe it never needs to be determined and the field of the virtual as infinite possibility is in itself enough). What is needed is an infinitely reconfigurable code that is non-representational, sonic, rather than visual.

Experimental music was also understood by Attali as having enacted a state of *depersonalization*. Without a dominant and meaningful code he said: 'The noise of matter, unformed, unsalable, confirms the negation of meaning' (114). Within experimental or theoretical music, he thought, improvisation and the false idea that everything was possible had replaced the centralized power of the orchestra and its conductor, as 'caricature of self management' in a repetitive society that was described as a 'simulacrum of nonpower'. 'In fact, the most formal order, the most precise and rigorous directing, are masked behind a system evocative of autonomy and chance' (114). The theoretical musician for Attali served to enforce this cloak of freedom that concealed a kind of tyranny that was likened to the preprogrammed option the synthesizer offered in the wrongly assumed democratic music of Philip Glass. 'The place of an individual in the modern economy is no different from that of Glass' interpreter: whatever he does, he is no more than an aleatory element in a statistical law' (115). Once more the question that I have is what is the problem with this? Attali calls this, in reference to Michel Serres, 'the voice as one of the things of the universe' (115). And again, I see no problem with that.

Attali described a situation where musicians rather than having become liberated, were indeed enslaved by their own ill-conceived commitment to freedom, chance and the power of technology. 'Music escapes from musicians'. He said: 'Music, like political economy- and once again far in advance of it – experiences the transcending of men by their knowledge and tools' (115). Attali was keen that 'men' and musicians specifically wrestle back control of both tools and knowledge. For if musicians were enslaved by a commitment to a false ideology, then so must we all be. This was an essentially humanist approach, in search of existential solutions to cosmological phenomena. He was alarmed by the extent to which scientism had excluded the human as part of a process of depersonalization. Others would later come celebrate this.

As well as enacting depersonalizaton experimental music also led to *elitism* according Attali's account. The inability of theoretical music to find an audience and the constant rewriting of its codes gave rise to a situation wherein such music could only ever be appreciated and 'understood' by a finite elite group. And furthermore, 'Accepting nonsense as the foundation of its power, it precludes any chance of its developing an acceptable rationale for its domination' (116). The claim was that it had become characterized by a self-serving pretentiousness that destroyed any use value in terms of meaningful communication.

So, for Attali, by 1977 the situation the world found itself in was this:

> A music without a market imposes itself on an international elite, which once again finds itself exceeding national cultural traditions, seeking the Esperanto it needs to function smoothly, to communicate effectively: the dream of achieving a world-wide unity of the great organisations through the language of music, a language that finds its legitimacy in science and imposes itself through technology. (116)

The modern musician, despite affecting the appearance of an independent artist, was in fact working at the behest of the great powers, and was the producer of the symbols of power. It was, for Attali, a power that confirmed the end of meaning.

Theoretical music's response to 'empty' pop music, which occasionally allowed for quality output, was not to Attali's liking. Rather it took repetition to a new level of absurdity. All that was meaningless in pop music of the 1960s and 1970s was emptied even further by the avant-garde modernists. Mass music and theoretical music were for him more tightly entwined than was immediately apparent. Both operated in the service of a dominant technocracy that had reached a critical turning point. Like most turning points, the future direction of travel involved a choice: In this case between a new kind of music with liberatory qualities and a newly encoded dictatorship. At its zenith, Attali believed, repetition imposed silence, and it did so by means of controlling noise. 'One must then no longer look for the political role of music in what it conveys, in its melodies or discourses, but in its very existence' (122). For music, in all its forms, in the mode of repetition was meaningless. Its very existence silenced noise. The mode of repetition demonstrated this never better than in the concert halls. For the elite it was an opportunity to remind themselves of their moral superiority, and for the rest it became a simulacrum of the festival, a representation of violence and release. The controlled environment of the concert hall had become a simulacrum of life. If an excess of life is death, then noise is life. In turn then too much noise, as another kind of excess, must mean death also, and this was identified by Attali as a feature of music, theoretical and mainstream, in the mode of repeating. What was required to overcome it, he thought, was a new language that could return meaning through new codes.[8] 'Today, the repetitive machine has produced silence, the centralised political control of speech, and, more generally, noise' (122). So, noise needed to be liberated, but only to a degree. It would need to be a controlled noise and coded in a way that did not induce further death and silence, but that could disturb the silence of repetition. For Attali repetition had prevailed to such an extent that under its conditions, and within the framework of its codes, no one had anything to say any more. Their voices could not be heard above the 'controlled noise' of the system – a system that had succeeded in employing even dissenting voices as a central aspect of its control mechanism.

For Attali then noise implied dissent and negation, and was a force that although essential needed to be coded and controlled. The quite different claim here is that it is a universal ground of potential. It names a process of incessant destruction and renewal that can only ever be tamed temporarily. Whenever a system processes elements that it deems to be dangerous, inconvenient or irrational, it prepares them for integration into that system. Such processing produces waste, and does not simply disappear, but is stockpiled, and as best as possible, safely stored and effectively silenced. Noise is returned when this waste can no longer be silently stockpiled. It is in these marginalized zones where the possibility of new forms of expression lies and rather than being silenced they should be listened to. Interestingly Attali also recognized this to some degree. Repetition, he said, stockpiled use-time. Things that used to be done like engaging in music as a temporal act had been objectified in artefacts like records that stored this experience. In such a technological form the experience was stored away, free of noise. Like money it solidified time in a set of relative values that did not take time to establish. Thus, in the mode of repetition music became the herald of death in so far as it removed the need to live. Noise was silenced, first by music and then finally by the commodification of music in technological form.

The complex nature of this repetitive economy where demand was produced prior to production and distribution, as the systemic management of noise, ultimately reached crisis point when that demand could no longer be sustained. When the mode of repetition found itself in crisis, it was evident in new ways. No longer was it a harmony torn asunder by dissonance, opposition, revolt or dialectical questioning, but the product of the proliferation of excess that could not be matched by the production of demand. This crisis could either be contained or allowed to proceed to its logical conclusion, Attali thought, whereupon the conditions would be appropriate for it to be succeeded naturally. On the one hand he was questioning the dialectical notion of succession while at the same time allowing it to creep back into his thinking. Ultimately, the resolution of the crisis was inevitable he thought. The market was expanding to the point where everyone had access to everything, even that which had previously been reserved for the consumption of the elite. Under these conditions he detected signs that the conditions were right for the emergence of a new kind of language. This condition was described as 'the proliferating circulation of pirated recordings, the multiplication of illegal radio stations, the diverted usage of monetary signs as a mode of communicating forbidden political messages – all of these things herald the invention of a radical subversion, a new mode of social structuring, communication that is not restricted to the elite of discourse' (132).

Composing

Attali was issuing a call to the audible as a site for critical engagement that could account for contemporary problems or opportunities. Such was the paucity of critical potential as it existed within the framework of political thinking in 1977 he thought, that any opportunity for progress had to be grasped. He heard such potential in emerging new musical practices that were beginning to pierce the silence that had been imposed on noise.

> Today, a new music is on the rise, one that can neither be expressed nor understood using the old tools, a music produced elsewhere and otherwise. It is not that music or the world have become incomprehensible: the concept of comprehension itself has changed; there has been a shift in the locus of the perception of things. (133)

The call for a new mode of thinking is clear, but where is the elsewhere that Attali speaks of, and what is the otherwise? These are largely dialectical constructs and require a shift from the visual to the audible so that we can reconstitute what we understand a 'thing' to be – a real thing, a virtual thing, a primary thing, a secondary thing, a visual spatially extended thing, or a sound thing/object. For Attali such a shift in thinking would surmount the crisis of repetition by imposing a new and different code. But there was little or no indication of what a new mode of perception might be, aside from an emphasis on listening. It is proposed here that the figure of noise, when read through the thought of Jean Luc Nancy, and drawing on the process philosophy of Alfred North Whitehead, can support such new and different modes of engagement that do not rely on the kind of dialectical logic that would produce nothing more than a contemporary noise reduction strategy.

Attali detected the early warning signs of a new kind of activity in music that was once more undermining the dominant codes of political economy. He went on to say,

> There is no communication possible between men any longer, now that the codes have been destroyed, including even the code of exchange in repetition. We are all condemned to silence – unless we create our own relation with the world and try to tie other people into the meaning we thus create. That is what composing is. Doing solely for the sake of doing, without trying artificially to recreate the old codes in order to reinsert communication into them. Inventing new codes, inventing the message at the same time as the language. Playing for ones' own pleasure, which alone can create the conditions for new communication. (134)

The existence of a code that determined the aims of the producer meant that they lost any genuine control over their labour. The exterior imposition of a code, in music as in other sectors, changed the dynamic of the creative process – essentially stripping it of meaning. Reasserting the 'authenticity' of motivation was then key for Attali in terms of making music purely for pleasure rather than it being enacted in the service of *scientism*. Music he thought should be an end in itself. In stark contrast to theoretical music that he said was guilty of creating langue at the same time as parole, new forms of musical language could instantaneously revive meaning, and reconstitute historical ambitions.

But where were the new codes? Not in Russolo, not in Cage, not in the Rolling Stones. They were not new but were rather 'the liquidation of the old' for Attali. Instead he thought,

> Free Jazz was the first attempt to express in economic terms the refusal of the cultural alienation inherent in repetition, to use music to build a new culture. (138)

Other forms of jazz had been appropriated by capital and repetition, it had become acceptable as part of an industry. Free jazz on the other hand had set out to establish a parallel industry to support its new mode of experimentation, often organized formally into associations and guilds (138). These groupings would pool royalties in order to support the creative process without the imposition of commercial pressure. Ultimately however, Attali suggested, it failed. It did succeed though in breaking down barriers between popular and learned music in its fusing of Black popular music and 'abstract theoretical explorations' of European music. 'But since this noise was not inscribed on the same level as the messages circulating in the network of repetition, it could not make itself heard. It was the herald of another kind of music, a mode of production outside repetition – after having failed as a takeover of power in repetitive society' (140).

Again, Attali concentrates on the formal characteristics of political economy. There was little if any assessment of free jazz as a musical form that was disruptive in terms of pattern formation, as the sound of discontent or as 'the noise of time'. Attali introduced the possibility of a new mode before retreating from it. These 'nearly events' are always heralds for him, and they emerge as parallel or abstract phenomena, both stylistically and structurally. Yet there remained a confidence that by 1977 something was stirring the tide of history; that another time was imminent. Yet it never seemed to arrive, and indeed it still has not arrived. But what if music was not a herald after all, but the resonating and noisy process that neither begins nor ends and which cannot be differentiated into finite periods? While less able to offer the predictive potential of Attali's model, this would offer a different way of understanding time, history and political economy – not as a succession of phases but as a perpetual and unpredictable reorientation: more Schoenberg than Adorno.

Because, 'Better than Adorno, Schoenberg understood the master dialectic of musical history, the back-and-forth between simplicity and complexity. "I cannot deny the possibility," Schoenberg once wrote, "that as often in the musical past, when harmony has developed to a certain high point, a change will occur which will bring with it entirely different and unexpected things"' (Ross 2012: 390).

As the master dialectic of musical history, or as that which removes the need for a master dialectic altogether, noise will always produce 'unexpected things'.

Attali then turned to another musical phenomenon to identify and underpin his phase of composition. This time it was the resurgence in popular music of traditional instruments. It was a resurgence of music he said for 'immediate enjoyment, for daily communication, rather than for a confined spectacle' (140). It was an accessible music that required no formal training and was largely improvisational. 'It has developed amongst all social classes, but in particular among those most oppressed (the workers of the big industrial cities, Black American ghettos, Jamaican shantytowns, Greek neighbourhoods etc' (140). Quoting O. Revault d' Allones, he describes it as a music of the disenfranchised, the foreigner, the other, the metics (140).

The 'mode of composing' was then characterized by amateur improvised orchestras playing for free and using traditional instruments. In some respects, what Attali described was a sonic economy of infinite potentiality. Rather than it being a utopian teleological end point, a successor to a previous mode of being, it can be read as a description of a *process* – one that operates throughout all of the preceding modes; a way of understanding not just the future but time itself as non-linear. Composition was described as a way of working with sounds in a way that was not constrained by grammatical structures that could direct its purpose or shape. It was a return to the festival spirit, not yet sure why or how it operated, and not subject to a singular guiding voice or set of principles. In short, it was everything that theoretical music was not. It was a realm of pure potential that stepped out of the technologically restrained system to pursue new relations. In essence it transcended the technological code that Heidegger had described as enframing (1977). It was a joyous return to a simpler more spiritual kind of music. It was a pure form of music that in his assessment once again appeared to be premonitory. Composition, he said, was characterized by a burgeoning of instruments that was as great as it had been in the sixteenth and seventeenth centuries, which had announced the industrial revolution. This time they foreshadowed 'a new mutation in technology' (144).

This mutation largely involved, or demanded, a human reappropriation of technology – or in the case of the music Attali described, a rejection of it – to create new forms of expression. Composition liberated time, he thought, so that it could once more be lived, not stockpiled: 'It is thus measured by magnitude of the time lived by men, which takes the place time stockpiled in commodities' (145). This would potentially usher in a new kind of political economy. It was a call to arms for man to finally embrace all previous historical advances, and to develop new forms of meaning and new systems of organization. Music, he said, 'no longer recounts a mastered, reasoned history. It is inscribed in a labyrinth, a time graph. After the third stage of the attainment of power described by Castaneda has been passed, the stage in which man has conquered power, the relation to technology and knowledge changes, because the relation to the essential has changed' (147). Attali's version of such an essential anthropocentrism assumed a transcendent set of practices that were capable of countering the dominant code of repeating. He assumed of music something special, and even perhaps that it was possessed of mystical powers. Such powers, he thought, were capable of allowing us, like with Heidegger's way of questioning, to access truth.

> Composition thus leads to a staggering conception of history, a history that is open, unstable, in which labour no longer advances accumulation, in which the object is no longer a stockpiling of lack, in which music effects a reappropriation of time and space. Time no longer [in the mode of composition] flows in a linear fashion; sometimes it crystallizes in stable codes in which everyone's composition is compatible, sometimes in a multifaceted time in which rhythms, styles and codes diverge, interdependencies become more burdensome, and rules dissolve. (147)

I take much from this idea of composing: The relational free formation of music as a model of universal patterning for instance. And the idea of a 'multifaceted time' that seems an apt description of the digital present, as has been argued by Virilio (2005) and more recently by Barker (2012). But unlike

Attali I see the potential for this to be realized in a post-human way. I do not fear for the loss of music and creativity to technology, or to scientism, but hear in music – all music – an energy, a noise, that can aid our understanding of chaos and complexity as noise, rather than serving as a means to control it. Music, and the new codes of composition are immanent rather than transcendent, not a solution but a feature of a technological phase in history wherein time moves in all directions simultaneously. For Attali, technology was an aspect of repetition, and the development of a purer form of music as composition signalled a volkisch recommitment to Being and to being human. Composition was counterposed in relation to repetition as dialectical herald. However, in the non-dialectical field of noise, technology is bound up in a nuanced and complex set of relations that move and vibrate at variable speeds, as relative to humanity, and not opposed to it.

Attali's work examined the cause and effect relationship between the political economy of music and political economy more generally. In doing so the role of political economy *within* music meant that political economy itself and not music was the focus. Yet it is clear Attali wanted to do something other than that in so far as he urged us to listen, to analyse *through* music, using a 'musical' methodology. But it was always unclear exactly what form such a methodology might take. What follows then is an analysis of music that is embedded in, and not resistant to, related technological landscapes and their temporal unfolding.

Notes

1 For a detailed discussion on the materiality of sound see: C. Cox, 'Beyond Representation and Signification: Toward a Sonic Materialism'. *Journal of Visual Culture*, vol. 10, no. 2 (2011): 145–161. For a counterpoint see, L. Döbereiner, 'How to Think Sound in Itself? Towards a Materialist Dialectic of Sound', Proceedings of the Electroacoustic Music Studies Network Conference Electroacoustic Music Beyond Performance, Berlin, June 2014.

2 But is there a time 'before political economy'? Should it be regarded as a named field of study or discipline, or is it a more basic human activity (or possibly non-human) that can be traced back beyond its having been named? William

F. Campbell sets out three distinct periods of political economy: new, old and ancient, and such a lineage casts doubt on Attali's assertions (http://www.mmisi.org/ir/12_02/campbell.pdf).

3 This loss of meaning leads to chaos and noise which relates to Nate Silver's work in *The Signal and the Noise* (pp. 118–122).

4 See http://www.bbc.co.uk/news/business-38341251.

5 For an oversight of the current state of the market in terms of quantity and quality see, Susy Frankel, Daniel Gervais, ed. *The Evolution and Equilibrium of Copyright in the Digital Age.*

6 Recent reports suggest that even this cynical kind of separation is being surmounted to create a flattened consumer base. What has happened to the musical generation gap? https://www.theguardian.com/spotify-family/2017/apr/26/whatever-happened-to-the-musical-generation-gap.

7 Interestingly Xenakis himself was sure that rather than acting as a herald, music had indeed lagged behind.

8 Richard Wagner made similar observations in his future music manifesto:

> In the shrill and frequent outcry of our shallow musical dilettanti for 'melody, mel-ody!' I find evidence that they take their idea of melody from musical works in which, in the place of melody there stretches an expanse of unmelodiousness, setting the melody they mean in the light they love so dearly. In the opera houses of Italy there gathered an audience which passed its evenings in amusement. Part of this amusement was formed by the music sung upon the stage, to which one listened from time to time during pauses in the conversation. During the conversation and visits paid from box to box, the music still went on, and with the same purpose as one assigns to table music at grand dinners, namely, to encourage by its noise the otherwise timid talk. (Wagner Music of the Future)

3

Remembering the Future: 1977–2017

The following is an account of the noise of time from where *Attali's* work ends. It is not a full or comprehensive account, and where the gaps are evident, readers will fill them in to further demonstrate the complex and chaotic nature of a pattern that can never be fully drawn or represented.[1] It would be impossible here or anywhere, to account for the entire history of music during the period being considered. Too many imponderables prevail: What is music, how do we categorize and measure it? More importantly how do we keep track of it given that it happens in so many places at so many times in so many different forms? What counts, what does not? What is significant and what is not? Answering such questions might serve to quieten the noise, filter the waste or compress the data, but the point here is to reside in the noise – to combine speculation with science (to rethink the human rather than removing the human altogether), and to describe how the process of listening, although limited in scope, is indicative of a different sensibility.

Much of the period under consideration has already been well-documented, but certain significant moments/fragments should be heard to get us to the present via a history of the future. Simon Reynolds in *Rip it Up and Start Again (2005)*, *Energy Flash (2008)*, *Bring the Noise (2007)* and *Retromania (2011)* tells a very detailed story, and his narrative has been augmented by a host of other commentators, critics and archivists, too great in number to acknowledge here. What follows then, rather than seeking to add to this body of knowledge, is an attempt to listen, and to do so within the theoretical framework argued for thus far. It will leap and stutter, pause too long in places and leave gaps that will perhaps infuriate the reader in others. But this is neither a logical continuity nor a discontinuity. On the contrary, it is a necessarily chaotic *listening to* – not by a subject/author who is in control of universal essences but by a resonating being who forms part of the pattern and who has in turn been patterned by, many of the instances/events outlined here (and who is happy to pattern and be patterned anew without foresight). Such an engagement with the universe through noise via listening means allowing oneself to be met by the world, embracing as a positive the richness of complexity, uncertainty and chaos. It recognizes the futility of forcing a will on a world that cannot be fully tamed, on a process that cannot be stemmed. With this approach comes an acknowledgement of potential, and a celebration of change and surprise. A journey through noise is a form of creative engagement. This is a brief diary of my own journey and an examination of the work of a number of artists who give us a sense of how it can usefully serve as a method of engagement.

Forty years have elapsed since Attali's seminal work was first published. It has been forty years of continuous discontinuity; forty years since hope sprung eternal in composition. Forty years since music seemed to herald another significant change. Forty years and much has changed. Forty years and much has stayed the same. The 'herald' was once again a false dawn it seems. What does that mean should be concluded in relation to Attali's work? Does it mean we wait – listening for the next herald? Wait and hope? What follows will implore that we stop hoping and disentangle the subtle intricacies of his text from the headlines, and work out what, if anything, noise, sound and music tell us about our current plight and its temporal dimensions. The examples chosen in the service of enacting this task are not meant to tell us anything other than there are dynamic resonances and dissonances at play that we may or may not encounter and which may or may not encounter us. To continue in this vein has political and economic consequences. It precludes historical destiny and requires a new or *post*-political economy that is speculative and curious, but also accepting of that which we have not yet encountered directly (as fifth dimension). Such an approach is orientated towards a constant engagement without settling, or only settling/pausing momentarily. There is always more work to do, more to find out, more mistakes to makes and more accidents waiting to happen. This is noise and although it may be possible to a degree to model some of its characteristics and patterns, it can never be fully silenced. To enact this new form of political economy requires listening in a way, which Attali elected not to do. It requires listening to specific instances of noise organized as music.

In his documentary *Hypernormalization*, Adam Curtis states that in 1975 counterculture washed its hands off politics. Citing Patti Smith as an example, he almost blames them for turning their gaze away. Is this what was happening? Or was it that a new kind of politics was forming – a politics of noise, one not based on dialectical opposition and the pursuit of an idealized future? If so, was it heraldic in the sense that Attali meant it? In the documentary Patti Smith is seen wandering through New York, reveling in rather than lamenting its decaying ruinous state. Such reveling invokes a multisensory form of engagement of the kind called for by Christopher L. Witmore. It is a sonic archaeology where the spatial and the temporal become entangled and where chaos and uncertainty reign. As the fixed visual–spatial dimension falls into ruin so artistic expression retreats, or advances into the realm of the sonic – not to oppose in a dialectical fashion but to celebrate the many different forms of perception that are required to navigate such a scene. But was this a turning point, a portent of things to come, a herald for a new kind of politics, or finally an acceptance that as the Sex Pistols would later claim, there was *No Future*? (http://www.bbc.co.uk/programmes/p04b183c).

From Wagner's vision of the future and the romantic version of music as sublime, up to the present day, come a great many examples of music's ability to engage with futurity as a feature of the modern world, and these will be situated here within a temporality of noise. In relation to this temporality, in his *The Music of the Future*, Robert Barry writes about the availability of an easily accessible digital music archive giving rise to a situation where the past could be acting as a drag on the future, as 'retromania' (Reynolds 2011) takes hold. It is just such a situation that Curtis describes in his documentary. The inability of the future to escape the gravitational pull of the past is described as follows:

> In the back of my mind, I could hear the voice of the writer Owen Hatherley, who once said to me, on the subject of pop music, 'a well informed observer from 1976, suddenly exposed to the most forward-looking music of 1996, would have encountered something completely alien.' A similarly clued-up time traveller leaping from 1996 to the present day, on the other hand,

> would probably appreciate several interesting developments, but they would recognise them as just that: more or less predictable extensions to a known field. (Barry 2017: 29)

This describes a relative temporality with periods of fast and slow, loud and quiet. It also alludes to a paucity of newness in contemporary terms. Like Barry I am not fully convinced by this. It is not that nothing new happens but that new and old, each relative to the other, exist in a different kind of temporality, or multitemporality (Barker 2012). In noise, newness and innovation exist as 'an always moving point on a dynamic continuum' (Barry 2017: 26) that can accommodate both the glacial and the rapid, fallow periods and explosive moments that form complex and difficult to predict patterns. Like the example earlier where the time-lapse between Dixieland and Psychedelia, and Psychedelia and the present was contrasted, this temporal paradox is interesting but not convincing. A time travelling Johnny Rotten, or Pete Shelley, would not have been overly surprised (although they may have been horrified) to hear the shiny electro pop of Babylon Zoo's *Spaceman*. It is not the case that the future has been cancelled, but that it must be understood within a less clearly defined linear framework.

In 1977, the Sex Pistols played at the Free Trade Hall in Manchester. Attendance at this concert became the stuff of legend as legions of would-be musicians later claimed to have been there.[2] The concert was organized by a group of students who would later go on to form Buzzcocks. They were Peter McNeish and Howard Trafford aka Pete Shelley and Howard Devoto. Within weeks Buzzcocks were fully formed, rehearsed, ready to reprise the event with themselves in support, and also ready to release their own self-funded record *Spiral Scratch*. They had resonated with the energy they witnessed that evening and they immersed themselves in it. But this resonance would quickly turn to dissonance as Howard Devoto felt himself at odds with the pattern he recognized as forming. Punk had always been an uneasy alliance between a kind of perceived intellectualism and raw power, of art school experimentalism and working class aggression. As such it was not unlike the opposing forces of composition and theoretical music described in Chapter 2. Such uneasy alliances are common tropes in accounts of social, political and economic formations, as well as in philosophical and scientific accounts of music. In many respects punk was the manifestation of the historical phase that Attali named composing in so far as it constituted an outpouring of fervent 'music for music's sake' that was open to anyone who wanted to participate. Such was the complexity at play in this instance, however that it was difficult to see how punk was ever anything other than a momentary intensity that would energize an array of orbiting constellations without itself forming a coherent pattern, let alone a herald. Indeed, it operated as an energy that created orbits around it, in turn orbiting other intensities in a dynamic fractal pattern where beginnings–ends cause–effect were subsumed in a vibrating cosmology – a sonic economy, in a temporality of noise.

> Devoto exemplified the secret truth of punk: it was a movement based in the rebellion of middle class-misfits as much as those mythical 'kids on the streets'. He was studying Philosophy and Literature when he met Pete Shelley at the Bolton Institute of Technology and formed Buzzcocks. Punk's own rhetoric though, suppressed the art school and undergraduate contribution, and amplified the imagery of tower blocks, urban deprivation and youth unemployment. (Reynolds 2005: 15)

At its best, punk was both of these things simultaneously. In reality, this was never really a secret and nor was it the 'truth'. Instead it was an unstable conglomeration of factors that seemed to resonate briefly before dispersing. Like the experimental formalism and the romantic movements that preceded it, and like Nietzsche's Apollonian and Dionysian dualism it was an attraction of forces – a resistance that repeated (Nietzsche 1967). It was noise moving to the audible range as sound and music. It was an ordering rather than an unleashing of chaos, an attempt to 'understand' through resistance as Jean Luc Nancy might have it (Nancy 2007). It was an expression of the interface between multiple events as noise moved to sound, music and a plethora of other expressive art forms.

The movement from 1977 to 1978 was particularly significant in terms of the identity crisis that emerged so quickly in the punk movement that had established itself in 1976. It was also significant in terms of listening for the herald. Like much of what has gone before in this book, dualisms seemed to prevail, as punk became polarized. But as always, the interesting and informative moments are to be found in the chaotic and noisy regions in between the poles – in the energized core that is the temporality of noise, as the passing of time and that which makes time pass. The struggle between willful experimentation and a more basic music for the 'people', as we discovered earlier in relation to Attali's critique of theoretical music, was a familiar one, and the failure to reach a resolution is what makes it so interesting.

Devoto rejected the standard view of punk that for him had become a tabloid cliché. Shelley and his commitment to a lineage of avant garde music from Iggy Pop to Krautrock had always positioned them at a specific point on the punk rock spectrum, and according to the music critic Paul Morley, Buzzcocks were akin to free jazz and the experimentalism of Ornette Coleman (Reynolds 2005: 16). Theirs was a resistance based on sonic disruption and experimentation, traversing time and space with superpositional qualities as it engaged with traditional rock tropes to address present issues, but in this case with a distinct orientation towards the future. 'Boredom was why Devoto decided to quit on the eve of Spiral Scratch's release. He was fed up with the "unrelenting nature" of punk music, including Buzzcocks ultra fast thrash. He told the press he was 'tired of noise and short of breath' (Reynolds 2005: 17). The noise annoyed him. He was driven by what he called 'negative drive'. 'In fact "negative drive" was always what I believed the punk ethic should have been about: constant change, avoidance of stale conceit, doing the unacceptable' (Reynolds 2005: 18).

In 1977, Devoto left Buzzcocks to form Magazine, allowing him to pursue his longing for 'Bowie-esque uniqueness' and fluidity that was, ironically, not at all unacceptable. Undeterred, Buzzcocks continued at full speed with their own form of expression that allowed them to engage with all aspects of their world without recourse to one or other of the perceived poles, pursuing themes from love and consumerism, to science fiction and technology. The sound, although guitar driven, always had another quality, an angular modernism that chimed with Shelley's own interest in computers and electronic music. He had built synths as a student and produced a body of electronic material as early as 1974. This work was rediscovered and released in 2016 with the titles, *Sky Yen*, and *Cinema Music and Wallpaper Sounds*, and is testament to the entanglement of past, present and future that has always been evident in his work, and to the strange temporal unfolding where continuous and discontinuous patterns can be discerned. Shelley's oeuvre is notable as much for a consistent lineage as it is for sudden ruptures and breaks.

Devoto's new direction was more explicitly 'intellectual'. He was unapologetic in his rejection of punk's constraining and generic characteristics. This split was an echo of previous splits as discussed earlier, between romanticism and formalism, and between intuition and intellect. Such dualisms generally imply a zero-sum choice: A this way or that way dilemma? But this constitutes a slightly narrow reading. It may be the case that certain very clear examples can be cited to illustrate the existence of an either–or choice, but there are many more interesting instances of indecision and uncertainty, of constant vacillations between two poles and the refusal to make a choice – not least Buzzcocks whose music defies definition and celebrates eclecticism and experimentation without pretension. The point here then is not to debate historical accuracies or assign actions to one category or another, but to identify instances, should any exist, of the herald that Attali believed would announce new codes and reconfigured forms of political economy as composing superseded repetition. If anyone encapsulated this joyous freedom of expression that characterized composing then surely it was Pete Shelley. Yet his and the music of those around him did not in retrospect herald any new dawn.

Howard Devoto, despite his unapologetic intellectualism, was also explicit in his refusal to be identified as belonging to either of the opposing tendencies. As such it could be argued that he too was an exemplar of the new code of composition that was rejecting all that had gone before. Referring to Magazine's first single *Shot by Both Sides* Simon Reynolds says,

> Without specifically referring to any of the great divisive issues of late seventies Britain (Rock Against Racism and the Anti Nazi League versus the resurgent far right; the collectivist left that was taking over the Labour Party versus the pro-entrepreneur right wing that dominated the Conservative Party), 'Shot' captures the era's sense of dreadful polarization, and the vacillation of those caught in the cross fire with the centre ground disappearing beneath their feet. It is about a non-combatant, an *in*activist. It's a defence of the bourgeois

> art-rock notion that the individual's struggle to be different is what really matters. (Reynolds 2005: 21)

If *Shot by Both Sides* was the sound of a country lurching between two extremes, then the noise that Devoto was tired of was the trembling of time as it collided with its own field of virtual possibility. Not a bifurcation of incompatible practices but a process of tearing and colliding – as noise. The sides that Devoto felt caught between did not necessitate a dialectically induced third space to retreat to, seek comfort in, in accordance with Attali's description of composing. Rather it was an intensity of noise, to be encountered as an energy around which new and different patterns were struggling to form. For Reynolds, the temporal chain breaks in 1977. From 1978 onwards, new lines of flight opened up. Led by pioneers like Devoto (but more importantly by Shelley according to the account being developed here) new possibilities were revealed as other more reactionary forces took 'conventional' punk in an entirely different direction. But Devoto was not alone. Other pioneers were also opening up new avenues. John Lydon (formerly Johnny Rotten of The Sex Pistols) was forming Public Image Limited, and Vic Godard was rallying against the rock n roll establishment with his band The Subway Sect.

> Where Howard Devoto deliberately cultivated intrigue with his oblique comments and evasions, Godard seemed simply to exude a haze of indeterminacy… They looked regular but slightly out of time. (Reynolds 2005: 25)

Musically too they were slightly at odds with the times, producing a sound that did not seem quite right. And things were not quite right. Godard had dismissed punk almost as soon as it had started. It had begun to adopt the rock n roll pomp he opposed, and was bereft of any meaningful strategy for change. Equally bereft were established political programmes, and Godard adopted an intellectual stance that aligned him to neither wing of the political spectrum. In contrast to Devoto however, his was not a defection to a third space, but a difficult to define expression of the fact that very little made sense anymore; that established norms and patterns could no longer account for the complexities that were beginning to unravel. Like Patti Smith's reveling in the decay of New York, Godard's free and inquisitive adoption of indeterminacy was a compelling political gesture. He and his band's aesthetic ambivalence, combined with their sonic ferocity, seemed to articulate the presence of noise as resistance, in accordance with Hainge's definition, and not as a dialectical opposition, despite the 'we oppose all rock n roll' mantra. Perhaps because of this, no one quite knew what to do with Vic Godard or the Subway Sect. It was never certain what they were. They did not dress like the rest and they did not want to be like the rest. Their manager Bernie Rhodes was reluctant to release their material and the world would have to wait thirty years for their early work to surface in the shape of the temporally dislocated *1978 Now*, released in 2007 and *1979 Now* released in 2014. 1978 had seen the band peak and their support slot on the Buzzcocks tour, and the release of their second single *Ambition*, stand now as significant events in the story of post-punk and its journey through indie to the present day. For a moment it seemed like 'another way' may be possible. Buzzcocks, Subway Sect and Public Image Limited (among others, again too numerous to mention) seemed to signal new possibilities. Maybe Attali was right after all. Maybe a phase of composing was emerging from the ashes of repetition.

The crisis in repetition that Attali had identified was becoming apparent as the established world order appeared to be crumbling. Political and economic stability was now being supplanted by precarity. In the United Kingdom, although the music played the country endured a winter of discontent as James Callaghan and his Labour Government went head to head with trade unions. Across the rest of the world full employment was becoming a thing of the past, as inflation and the value of currency began to vacillate at alarming rates. According to one study, 1978 was a significant turning point. It represented a peak in terms of economic growth and well-being that would never be repeated. From this point on the world would be a very different place.[3]

But what if anything was the accompanying music alluding to? Did it contain any indications of what this different place would be like? If Devoto occupied a clearing on the outside in which it was possible to develop the codes necessary to access a new truth and to reinstate meaning, then Godard, Buzzcocks and PIL were adrift in the noisy zone of undecidability where resonance, dissonance and listening occur without the need for the imposition of a new code.[4] Theirs was less a dialectical strategy for change, and more an affirmation of dynamic energized potential. They did not constitute a pure form of musical practice or composing, not a blueprint for a new mode of political economy, but an expression through noise of incessant renewal and indeterminacy. This incessant deforming and reforming was evident as the political and economic struggles of the 1970s eventually resolved to harmonize. At the turn of the decade the country was beginning to show signs of a newly developing pattern as a distinct neo-liberal entrepreneurial spirit took hold. Composing seemed increasingly unlikely to succeed repetition in the way Attali had hoped. In fact if anything, repetition seemed to have entrenched even further, and would continue to do so up to the present. So, it might be concluded that music identified as being characteristic of composing did not herald. Instead it served a much more significant function, and this is the central claim of this book. Rather than being its antecedent music was entangled with political economy and it was the sound of history moving through noise – not in a uniform direction towards a predictable end but of a morphing where the future was not necessarily consigned to the past, as Fukuyama had thought, but was reconstituted in terms of its relationship *to* the past and the present. As such, music provided and continues to provide an analogical model for engaging with new political and economic formations without conforming to dialectical logic. It is the expression of the temporality of noise as a *post*-political economy, understood as the mode that includes all of the factors that traditional political economy largely ignores. The need to seriously take account of such a model is slowly being recognized, if not adhered to. The black economy, the dark net, the pink pound, the grey pound, minority literature, alternative music, diverse voices and perhaps most significantly the spectre of radicalism in all its forms. All are increasingly recognized as features of dynamic multifaceted systems where elements combine in an indeterminate fashion that resists compression and adherence to normative values. They can no longer simply be marginalized. No attempt to silence or ignore them is sufficient. The only option is to open up closed models of analysis and to replace them with appropriately agile and dynamic constellations where noise is reclaimed as a figure that is central to our critical functions rather than a negative descriptor of the peripheral and the unwanted. If the music of 1977–1978 heralded anything then it was this contemporary state of affairs.

Returning to the music: In the city of Sheffield during this period The Future was being formed (literally). In Attali's terms there was a 'new mutation in technology'. But here punk took a slightly different form, one that largely rejected the aesthetic that was coming out of London.

> Sheffield's preference for electronic sounds [also] related to a local spirit of futurism and technophilia shaped by the city's role as one of the engines of the industrial revolution. (Reynolds 2005: 150)

The grey landscapes that formed its urban centre, and the constant noise of heavy machinery that filled the air were regularly referenced as being significant by future stalwarts of Sheffield's music scene, Richard H, Kirk of Cabaret Voltaire and Martyn Ware of The Human League. Sheffield had a reputation for resolute leftism, and for people like Kirk and Ware being an artist was a way of establishing their radical credentials at a time when radicalism itself was fairly commonplace. In this case being an artist meant being receptive to ones' environment, local and global, and the technological changes that were beginning to emerge.[5] Cabaret Voltaire had formed in 1973 as what Kirk described as a 'garage band music concrete', a collective of artists and various other participants interested in film, literature and Dadaism. By 1974, the line-up had settled down to include Kirk, Chris Watson and Stephen Mallinder. The trio created soundscapes using a variety of experimental electronic and tape techniques. Like Shelley in Manchester, Watson had also built an early synthesizer, and this is yet another reason why time travelling inhabitants of 1970s Britain would not have been too shocked by any future sounds they might encounter 20 years hence.

Cabaret Voltaire and many other glam experimentalists in Sheffield embraced punk as their natural inheritance. But by the summer of 1977 its appeal was already beginning to wane. Ian Craig Marsh, Martyn Ware and Adi Newton had by now already formed their band The Future with the stated ambition to 'destroy rock n roll' – a desire that echoed similar sentiments to those issued by Vic Godard. During that summer, two records were released that would convince The Future that their ambitions were realistic. They were Kraftwerk's Trans Europe Express, and Donna Summer's I feel Love – both well-documented in terms of their status as *future sounds*. They collected an array of electronic instruments that although limited, allowed them to pursue new codes of expression. Described as a kind of enforced minimalism they persevered with their brand of melodic machine music. The interface between man and machine became central to their output.

> Initially, The Future came up with the 'rather radical idea that we'd have shared vocals' says Marsh. 'We dispensed with our names and called ourselves A,B, and C. It was all very computer orientated and linked to the lyrical composition program we created called CARLOS: Cyclical and Random Lyric Organisation System.' A cybernetic version of Burroughs and Gysin's cut ups or Surrealist automatic writing. (Reynolds 2005: 161)

The Future eventually abandoned this approach before finally parting company with Newton and carrying on making instrumental music, like 'Dancevision' which as Reynolds points out formed the blueprint for Detroit techno, another musical form orientated towards the future. Eventually they recruited Phil Oakey on vocals and changed their name to The Human League – going on to become one of the biggest bands of the 1980s and entering the 'mainstream' blending experimental sensibilities with Euro pop and disco. Such blending is key to the notion of movement and incessant combinatorics as a feature of noise and the always forming and reforming nature of music as a social practice that is in turn a feature of an always-forming universe. Not necessarily a specific herald, but always a model of renewal where outcomes are mostly uncertain.

The sound of Britain in 1981–1982 was a new wave sonic economy – a temporal soundclash with new codes and combinations of codes forming, not as a model to be imposed but as a complex and unpredictable system. While The Specials were capturing the spirit of the times in Coventry with *Ghost Town* – the sound of the present – Cabaret Voltaire claimed to be capturing something quite different. Not an explicit representation of political unrest, but a resonating sense that everything was about to change. Tempting as it is to cite such experimental music as a portent to wider technological innovation, it is taken here as a sonic expression of the present as large-scale industrial practices morphed into new form of economic activity, often characterized by the adoption of information technology.

The apparent revolutionary moment of punk then had arguably been little more than an indulgent art school prank, with a cursory reference to situationism, but its impact undoubtedly resonated outward as the decade closed. Music was telling us something about the present, and it was pointing towards the future. Its economic models were once again shifting, as they had before and would continue to do. But did it herald? Ultimately, I would say not. For if it had, then by my interpretation, an experimental, anti-establishment, democratic future lay ahead – a future where anyone who wanted to could do what they wanted to do. And they would do so in an egalitarian space of mutual respect and recognition of the rights of others, where encounters with the world were regarded as opportunities, or as Whitehead had put it, a potential *Adventure of Ideas*. But sadly this was not the case.

As the 1980s unfolded new and different patterns did begin to form. But instead of the idealized mode of composing that Attali had hoped for, came a reality that was, according to some, considerably less inspiring. The joyous and democratic music of the 1970s rather than providing the necessary code for a renewed sense of meaning through authenticity brought a relative degree of stasis. Contemporary research into the music of the decade, as well as providing the potential means to empirically measure Attali's predictive model, paints a fairly bleak picture. It suggests that mainstream music despite having remained stylistically diverse over a period of 50 years, demonstrated one particular exception: the 1980s. The research published in the journal *Royal Society Open Science*, developed an algorithm to map the levels of diversity and innovation in music in an attempt to identify both periods of continuity and moments of rupture or 'musical revolutions'. The researchers Matthias Mauch, Robert M. MacCallum, Mark Levy and Armand M. Leroi used the emergence of large, digitized collections of audio recordings, musical scores and lyrics to investigate the history of popular music from an evolutionary perspective. Drawing inspiration from studies of organic and cultural evolution, they developed a history of pop music as a 'fossil record' and asked the kinds of questions that a palaeontologist might, not unlike the questions Christopher L. Witmore asked earlier. They identified 1964, 1983 and 1991, as revolutionary moments in terms of quantifiable creativity. But what if anything did these moments communicate with regard to wider political and economic shifts?

Moving forward from these specific dates it is not immediately apparent that they heralded anything in particular. If anything, they signalled these period's sameness rather than change and there was no significant or radical advance of the kind Attali described. But this research had no particular ambitions in relation to the political economy.

'The work is far and away the most comprehensive and sophisticated analysis yet of popular music', said University of Reading evolutionary biologist Mark Pagel, who studies trends in human culture and was not involved in the research. 'Many commentators attempt to link eras of pop music to social and political changes, but this program does not rely on preconceptions. Rather it allows patterns to emerge from [musical] data' (http://www.pbs.org/newshour/updates/computer-scientists-prove-80s-music-boring/).

As such it retreats from dialectical notions of linearity and causality, and measures the complex emergence of patterns without subjective infringement. As a refined recognizer of dynamic pattern formation, it can stand as not only a useful tool in its own right but as an adaptable model for understanding the wider aspects of political economy as themselves operating in a musical way. From this perspective, music is not perceived as existing in relation *to* political economy but only as a pattern emerging from data. Such emerging patterns are understood here however, as being linked or entangled with an infinite number of other patterns within a temporality of noise as a sonic economy. In some respects, this incessant and infinite patterning seems to support Atali's bleak assessment of a world in the mode of repetition, although at the same time negating his belief in composing. But it also raises opportunities for understanding the world as patterned from data and the dynamic interaction of fundamental elements that surmounts the problems that are encountered by both correlationism and phenomenology. It supports ultimately a move towards an algorithmic universe that can be reduced to mathematical rules. This would support in turn a Bachelardian discontinuous universe composed of discrete units, but as has been discussed, this cannot account for the chaotic unpredictability of more continuous forms of engagement and creativity.

If the 1980s were ultimately predictable on the pop surface, and I am not entirely convinced this was the case, there was as ever an undercurrent of creative energy: Sonic Youth, The Smiths, New Order, Weekend, Everything

But The Girl, The Fire Engines, Aztec Camera, Josef K and Orange Juice, Mo-dettes, The Au Pairs, Girls at our Best and many more that I have forgotten or regretfully never encountered (but was no doubt influenced by, even if indirectly). And in another more 'experimental' direction aligned to the trajectory of Cabaret Voltaire came a host of musicians and performers who had been influenced by Krautrock as a general point of reference and by Throbbing Gristle more specifically. These included Coil, Nurse with Wound, Psychic TV, Einsturtzende Neubauten, SPK and again many more. There was also the beginning of electro that would later spark the dance music revolution, and at the end of the decade, the Stone Roses and the second summer of love would preface the Acid House phenomenon that would run until the mid-1990s. Along with all of that was the beginning of subgenres like Shoegaze: with My Bloody Valentine, Lush, Slowdive and Slint that would later provide the impetus for post-rock,[6] not to mention hip-hop. In short, endless twists and turns and events that sometimes leapt into the future, often delved into the past, but always energizing a present that was destined to fade.

The sonic economy as it was emerging in the 1970s and 1980s was not confined to the areas or genres that have been discussed thus far. I acknowledge my own bias in this respect, as having encountered very specific situations in terms of philosophy and music, but a more global noise, sound and music should also be acknowledged. 'World music' became a specific category during this period. It was an important aspect of the unfolding global political economy of the time and was indicative of the commodification of music for sale in carefully manipulated markets where demand, as Attali has said, was crafted in the interests of capital.[7] It thus occupied a space between repetition and, in so far as it was an example of a return to purism and traditional means, composition. Afrobeat specifically had a moment of intense audibility in the 1980s, and has intermittently made an impact in the United Kingdom ever since – moving through the early proto garage scene of Soul to Soul, Acid Jazz and more recently to Grime (https://www.theguardian.com/music/2012/jan/19/the-rise-of-afrobeats). But just as world music was being packaged for potential new consumers, so too was Western music being adopted in areas across the globe. So once again it is not easy to identify where, if at all, Attali's category of composing was happening or indeed what it was he had in mind when describing it.

Into the 1990s and Cool Britannia aside there was about to be a technological–techno revolution. The digital was coming but was it heard first? Turntables began to outsell guitars, as dance music came to the fore. Digitally produced records played on analogue decks prevailed, without a live band in sight. The recorded form had triumphed, but in a peculiar way where DJs were themselves 'live' performers – not always skilled technicians like the hip-hop turntablists, but people with 'taste', in a world where knowing what records to buy and play became an art form rather than a simple consumer activity. This was an interesting phenomenon in terms of the status of objects within an economy of repetition. Objects, in this case records, were meaningful and present in terms of the interplay among them across two decks, as part of an interaction where the movement of one into the other became more significant than each on its own. It was the moment of resonance, of passing one from the other that meant something and which drew towards a further encounter with listeners who were attuned to the transitions in the mix. Thus, the status of object did not mean an internalized stockpiling, but an interactive communication in a universe where every object somehow connects with every other. The appropriation of the record deck and its augmentation with a range of technologies that were emerging at the time became manifested in a number of new forms: hip-hop, house and techno. The latter emerged as a coming together of British experimentalism, Krautrock and the free jazz expressionism of Sun Ra that would later morph into Afrofuturism. What if anything though did it herald? Deindustrialization, a descent into ruins, post-industrial neo-liberalism, globalization? Like punk before it was most likely not a herald at all. Instead it was a sonification of events as they were being processed.

A key exponent of this new musical form was Rhythm is Rhythm (Derek May). Alvin Toffler described May as 'providing the soundtrack to an alternative future – where the people reclaimed technology for the benefit of the community'.[8] That alternative future was embraced in the United Kingdom by a generation of people for whom partying was a way of life not a weekend treat. Again, this story has been well-told but a short précis will be useful.

> In Ibiza in the summer of 1987 celebrating Paul Oakenfold's 24th birthday, Johnny Walker, Danny Rampling, Nicky Holloway and Oakenfold himself had a collective epiphany at Amnesia, hearing Balearic DJ legend Alfredo under the influence of ecstasy. They came back determined to recreate what they had experienced in their own style. Oakenfold started playing house at his Project club, then started Future, then Spectrum at Richard Branson's club Heaven. Rampling and his wife, Jenni, started Shoom, in a fitness centre in Southwark, now close to Tate Modern, but then a cultural desert. (https://www.theguardian.com/music/2014/feb/23/acid-house-dawn-rave-new-world)

Terry Farley, alongside Andrew Weatherall, Cymon Eckel, Steve Mayes, Pete Heller Started Boy's Own, another significant collective that momentarily signposted a new sensibility and potential for change (https://www.vice.com/en_uk/article/how-boys-own-changed-british-dance-music). The old codes were being rewritten and it seemed as though a form of composition had begun to establish itself. For a while the do it yourself (DIY) mentality and total commitment to music as a lifestyle seemed to offer an alternative to meaningless repetition – or a reappropriation of meaningless repetition elevated to a higher form as a cunning dialectical intervention. Either way it cannot be reflected upon now as a herald of composing.

In the United States in the 1990s, a second wave of techno was beginning to emerge spearheaded by artists such as Underground Resistance and the associated Drexciya. Both forged their own brand of futurism as the sound of technology combined with chronic urban decay and social deprivation. For some it sounded dystopian but for others it was joyous and liberating, containing tales of subterranean species that were forming new worlds, discovering lost worlds and promising of worlds to come. Like Sun Ra before them this other-wordliness was significant as both a metaphor and the realization of a future arrived at in the present.

'The lynchpin of legendary group Drexciya, alongside partner in crime James Stinson, Gerald Donald's uniquely experimental attitude has inspired a generation of producers. Working under a number of aliases and reluctant to give interviews, Donald remains remarkably mystified, operating in an orbit of his own. Just as his music seems to escape any sense of time and space, so too does his identity' (http://thevinylfactory.com/features/demystifying-gerald-donald-the-man-behind-drexciya-and-dopplereffekt/).

This escaping of time and space is significant. The post-human soundscapes they produced did not fit with the successive periodization of history that formal dialectical models rely on. Instead they referenced and echoed a constantly occurring process of complex pattern formation, within which humans are but a part. Within such complex environments, it is difficult to conceive of music as preceding in the linear sense. Instead it is a model for and an expression of, combinatorics, of renewal and decay as noise.

More contemporary examples of such complex pattern formation within this temporality of noise include: Howlround, JLin, Klara Lewis, Nils Frahm, Mary Orcher, Katie Gately, Pye Corner Audio, Moor Mother, Felicia Atkinson and Jefre Cantu-Ledesma, Max Richter, Oneohtrix Point Never, Johann Johannsson, Holly Herndon and Laurel Halo (once again among many others). These artists in one way or another, engaged with questions of noise and temporality. Whether through electronic experimentalism or classical orchestration, and in most cases both, they meet the world and are met by it in a manner that allows for a kind of non-representational expression that interfaces between the analogue and the digital, between the organic and the inorganic, between human and machine. They produce *future sounds* that call for a multi-temporal engagement with our time, a time of noise, where listening is more than a singular sensory activity, and is a form of engagement that is essential in identifying patterns, unpredictable as they may be, that emerge in our digital present – a present without causal past or teleological end. Some of the music cited might be described as theoretical, experimental and in some sense then, meaningless. Other examples are more inclined towards representation and traditional musical forms. Either way no distinction is made here, as in noise they constitute the complex and rich expression of the patterns forming and deforming through a constant interplay of dissonance and harmony.

A specific example is Oneohtrix Point Never's (aka Daniel Lopatin) *Garden of Delete*. Working through the simulacra that is Lopatin's virtual alias Kaoss Edge, it sonifies the fractured discontinuous environment and dystopian nature of the digital present. It is the sound of entanglement. As the album unfolds the world sounds like it is breaking up on a continuous/discontinuous basis, forming a *sticky drama* of temporal dislocation where the constituent parts are struggling to extricate themselves from the patterning tendency of organized repetition.

> *Garden of Delete* is an electronic album, but it's also the most rock thing Lopatin's ever done. The hypergrunge movement of Kaoss Edge might be fiction, but it accurately depicts the synthesis of cybernetics and heavy metal that makes up the record. It's full of lurid electronic presets that sound like a guitar blasting out of a wall of amplifiers and palm-muted note runs that sound like painstakingly sequenced MIDI, a grotesque, sinewy collection of sounds that evokes the intertwined sensation of curiosity and disgust I felt browsing the horror section of my local video rental store as a child in the early 1990s. It revels in its juxtaposition of intense, pallid noise and moments of chilling pin-drop calm, and if it was a movie, it would meld the body horror of David Cronenberg's *The Fly* with the bad taste schlock of *Leprechaun*. (http://www.factmag.com/2015/11/12/oneohtrix-point-never-garden-of-delete-interview/)[9]

The record also sought an entirely new approach to marketing, and as such it directly engages with political economy in its contemporary guise (or possibly even as an example of '*post*-political economy'). Daniel Lpoatin invented the imaginary band Kaoss Edge to promote the album. This caused a flurry of online activity of a kind that breached the defences of the real and the fictitious. Kaoss Edge became real, and they did so in a manner reminiscent of Jules Romain's *Donogo Tonka* (2009).[10] In this book, a strange temporality is formed where the simulacra precedes in a way that also echoes the temporal disjuncture set out in Lyotard's *Soundproof Room* (2002). A fictitious scenario is made manifest to protect the integrity of imagination. In both examples, what is created is not a depthless postmodern representational surface that can be linguistically deconstructed to reveal the truth. Instead it is its own truth, and this has significant political implications. This notion of a mixed reality without ultimate truth takes its place in the history–temporality of thought pertaining to the nature of reality, consciousness, intention, intuition, dialectical methods and the proposition of a rational political programme. Within such an economy rational political programmes that serve to bring the world to order, and represent it through the imposition of commonly agreed codes may no longer be possible. The imposition of order constituted in this way always creates waste products formed of material that does not fit the rationale. In a digital *post*-political economy, the borders between real and virtual, wanted and unwanted, perceptible and imperceptible are porous. The unwanted waste returns as everything sounds and resounds as noise and contemporary political thinking needs to embrace this. What Attali writes off as meaninglessness in a mode of repetition can then be rethought through noise to create new opportunities without recourse to composing as a newly compressed code. *The Garden of Delete*, sometimes incoherently, speaks to this contemporary reality.

In stark contrast to Lopatin's work is Max Richter's *Sleep*.[11] Coming as it does from an entirely different musical vernacular, it should be stressed that the point is not to directly compare the two works but to demonstrate how the perceived maelstrom of the digital present and its future trajectory can pass through different filters, collide with different traditions and aesthetic frameworks, to produce complex variations. In *Sleep* Richter cites the need to stop and reflect, to rediscover the rhythm of humanity. But rather than being distinct aspects of yet another dualism, between frantic activity and rest, these two very different works are closely connected as objects. On one level they speak to very different scenarios: The chaotic digital vortex of Lopatin, and the serene remembering of human vulnerability of Richter. On the one hand a toxic and violent swamp of primitive being as 'scientific image', on the other a 'manifest image' of calm after the storm. But given that the intention here is to take note of all music, inclusive as noise, the two are not regarded as mutually exclusive, not dialectical in their relationship. Instead the two connect in ways that reflect the earlier discussion of Wilfrid Sellars' thinking, and are also relevant to the more recent work of Steven Shaviro (2014) and his critical juxtaposition of the ideas Alfred North Whitehead and Graham Harman. Working with ideas that are temporally dislocated, Shaviro weaves the two contrasting positions together to create a compelling argument. He states that for the contemporary exponent of an object-oriented speculative realism, Harman, things–objects are almost entirely disconnected, internalized and self-sufficient, liaising with other objects only occasionally and vicariously. For

Whitehead on the other hand everything is connected to everything else as a cosmic orchestration of incessant renewal. Similarly, Oneohtrix Point Never and Max Richter connect as energized aspects of a single system of noise, where speeding up and slowing down, the rapid and the glacial once more, the Dionysian and the Apollonian, are not absolute states but are relative each to the other.

Through a critique of Harman and of speculative realism more generally, Shaviro resolves to foreground the idea of connectivity. Although recognizing the contributions made by the speculative theorists in terms of their realignment of human beings within the hierarchy of the Universe, and their reassignment as objects, he does not regard these or any other objects as essentially distinct or separate. For Shaviro it is Whitehead and not Harman whose position can best account for the contemporary present where connectivity is ever more apparent, and where momentary respite from its incessancy may be only fleetingly possible.

> Far from seeing any metaphysical problem of occasionalism or vicarious causation, therefore, I can only wish that some of the causations that continually beset me were indeed vicarious and occasional – instead of being all too overbearingly efficacious. For me, then, the great metaphysical problem is how to get away from these ubiquitous relations, at least in part, in order to find a tiny bit of breathing room. It is only by escaping from these overdetermined relations, by finding a space that is open for decision, that I may ever hope to find either Adventure or Peace. (Shaviro 2014: 34)

Even the retreat into a quiet space where there is breathing room, and sleep, is a feature of noise. Noise is always a relative interplay of rhythm and tempo. It is the dynamic interaction of everything in the universe audible and inaudible. The idea of 'speeding up' and the triumph of time over space, as discussed earlier in relation to Virilio, is a central theme here and Richter's work engages with it at an existential level. There is something very human at play in his work, yet equally it requires that we take account of a world shared with other actants, machines and processes that intervene with our natural rhythms and us with theirs.

Both Lopatin and Richter have something to say about the post-human, and the need to rearticulate our relationship to the contemporary digital age.

The former imagines a frantic staccato of glitch and error, an almost ceaseless dissonance, that only sometimes resolves to calm, although the latter deploys a more continuous series of variations and patterns that invoke somnambulant interaction with the universe. They both make audible our relationships with the rhythms, not just of ourselves and between each other, but also with the world to which we are connected. Listening to them supposes a kind of Being-in-the-world. In this case though it is a Being that is not essentially human, but a resonating with other beings and things as vibrating energy that moves at various tempos.

A further example of expressive practice that emerges through listening and then proceeds to engage with a range of contemporary questions as they relate to technology, political economy, the natural world and its ability to return order to the chaos, is the work of Johann Johannsson. On his recent soundtrack for the film *Arrival*, Johannsson takes sound as a cosmological force that is beyond human. It is wordless sound, akin to Barthes' *Grain of The Voice*, without meaning. By overlaying tape loops he builds sound until certain resonances emerge, not entirely by accident but in a way that allows the unexpected and the unpredictable to form. This technique or practice relates to the political gesture of listening where the world meets you and you consent to be met, to create an event, where 'the event (when thought through whitehead and deleuze), can be understood as a collection of processes, some of which are generated by a "user" and some generated by a technological system' (Barker 2012: 7).

The interface between user and system, meeting and being met, are consistent features of Johannson's work. In his *IBM 1401: A Users Manual* (2006), he meets and is met by, an object.

> Great ideas rarely spring from nowhere and the genesis of *IBM 1401: A User's Manual* is no exception. The lengthy evolution of this piece began in 2001 when Jóhannsson's father told him about the 'funeral' that he and his IBM coworkers arranged for the 1401 when it was discontinued in 1971. While he was obviously deeply attached to the computer because he was the primary maintenance engineer for the project (the 1401 was first mass-produced, reasonably priced business computer), the connection between Jóhann's father and his work actually extended into much deeper territory. Like his

> son, Jóhann Gunnarsson was an ingenious and musically savvy fellow, and he managed to figure out a convoluted way to make his computer 'sing.' In fact, the 1401 sang its own elegy at the ceremony, a brief theme from an old Icelandic hymn. It is a 30 year old reel-to-reel recording of this improbably sad ceremony that provides the central melody of the album's opening piece as well as the cornerstone of the entire endeavor. (http://brainwashed.com/index.php?option=com_content&task=view&id=8081&Itemid=1)

In this piece, Johannsson embarks on a research project to be expressed and realized through sound and orchestration, as the interface between man and technology is examined against the specific backdrop of Iceland's early adoption of computers. The computer as object and Johannson's encounter with it can in turn be thought once more through a number of philosophical ideas. In his *The Universe of Things* (2014), Shaviro tells us that for Graham Harman, through his category tool-being, objects are not merely delineated via recourse to theory or practice, but are instead entirely removed and withdrawn from human experience of them (45–64). Everyone and everything is unique, and carries within themselves characteristics that can never be fully accessed by other things.

> Everyone [and everything] carries a room about inside them. This fact can be proved by means of the sense of hearing. If someone walks fast and one pricks up one's ears and listens, say at night, when everything round about is quiet, one hears, for instance, the rattling of a mirror not quite firmly fastened to the wall. (From Franz Kafka's The Blue Octavo Notebooks and Max Richter's Blue Notebooks)[12]

Harman's thesis is a radical enactment and repurposing of Heidegger's work. For Heidegger, the *present-at-hand* description of objects as tools had a double sense: It accounted for the object that was largely taken for granted and hidden until, through a process of referentiality, we required and took notice of it, and yet it also revealed itself as existing at a deeper level when, in the case of a broken tool, it accentuated its being and in its singular uniqueness required that we take more serious note of it. This describes well the 1401, which Johannsson took note of at the point of its demise, when it was no longer fit for purpose. In so doing, the *present-at-hand* status of the machine revealed itself as having an existence beyond the delineable qualities of its generic suitability for use. Harman begins his assessment of objects by describing how Heidegger offered us a way of understanding objects in their own right, as having a life of their own, but it was by his own hand, not Heidegger's that this life was fully endowed. Things for Harman are not only significant when we require them, but have a life even when we do not know they are there, when we are not present to validate them. In short, the world does not depend on us.

> He [Harman] argues instead that the object's withdrawal from presence is a retreat from referentiality as well. This means the that 'the tool-being of a thing exists in a vacuum-sealed isolation, exceeding the relations that might touch it.' Instead of swinging between excess of referentiality on the one hand and an excess of singularity on the other, each object both disappears and emerges out of its own inaccessible vacuum. Harman carefully notes that, as a result of this reformulation, 'both Heidegger and Whitehead become direct opponents of my theory'. (Shaviro 2014: 51)

Heidegger would not have accepted the inaccessibility of essence, and Whitehead would not have recognized a world where objects existed in total isolation one from the other. There is then a highly or hyper localized dimension to Harman's thought. It implies a retreat into silence. For Harman things exist in a vacuum of their own being – inaccessible to all other objects/beings, except for in extraordinary situations when they meet/connect/interact. Reality is generally static with distinct substances removed from one another – silent. And this presents us with a temporal problem: How can change and renewal be accounted for within such a theoretical framework? For Harman (and for others such as Levi Bryant) newness and change is possible only because the essence of a thing is never exhausted by another knowing thing. It always has something new to give, something in reserve, something unexpected. For Whitehead, on the other hand, newness and change were brought about by constant interaction with things arising out of interrelationships that are dynamic, never complete and always in process – noisy.

> For Whitehead, echoing James, 'we find ourselves in a buzzing world, amid a democracy of fellow creatures'. Such a world is no longer human-centred: this is what unites Whitehead and Harman and the other speculative realists. In

> addition, such a world is one of discrete, individual entities, self-creating and self-subsisting to the extent that 'every component which is determinable is internally determined': this unites Whitehead with Harman's object-orientated approach, as opposed to other varieties of speculative realism. But the world envisioned by Whitehead is 'perpetually perishing'; thereby it also promises a radically open future. And this is what divides Whitehead from Harman. Where Whitehead insists on both internal decision and external relation, Harman has room for neither. And where Whitehead is concerned with both transience and futurity (which he calls 'creative advance'), Harman shows little interest in either of these. (Shaviro 2014: 40)

If, as Harman believes, objects are locked away in a vacuum (a soundproof room) they are in a way preserved in time – cryogenically frozen, archived, or maybe just abandoned. They withdraw into solitude until occasional encounters facilitate their vicarious mediation, creating two modes of temporality: An objective, and generally static universal mode (the scientific image), and an occasional and vicarious mode that constitutes lived experience and history (the manifest image), and which in the greater scheme of things, for Harman, is largely insignificant. This second mode is the linear temporality of Jaques Attali, a mode in which humans control noise, and in doing so, set in motion ordered historical advance. For Harman, and for others such as Quentin Meillasoux, the two modes of time coexist as a temporal dualism that reintroduces the idea of the world as it is and the world as it is for us. For Whitehead though, objects combine, constantly interact and vibrate towards their ultimate perishing, descending into ruin before their eventual rebirth in the temporality of noise. This creates a non-linear multi-temporality of relative tempos and rhythms that are always clashing, forming infinite patterns without defined beginnings or ends.

In *Soundproof Room* (2001), as a means of understanding this multi-temporality without recourse to beginnings or ends, Jean Francois Lyotard set out a version of time as it had been conceived of in modernism. He explained how the story and the narrator had been united, and how in the process the forces of nature and the universe bow down to human interpretation.[13] This account can be applied in varying degrees to the ideas of Sellars, as discussed earlier in relation to his stereoscopic model, to Heidegger, and to an extent to Harman, wherein respectively the dual aspects of temporality either combined or separate neatly. For Lyotard however, as it had been for Whitehead, they neither converge completely nor separate willingly. Rather they clash uncontrollably. The universe for them was not neat and could not be partitioned into ordered segments. To do so meant undermining another kind of time, a time of noise. This was not a time of successive ends and beginnings; nor however was it a time of death and despair:

> Rather than be extinguished by the languid swarming of the creatures in the dungeons where gods and heroes rot along with the rest, life is generated there. Death, be it chosen – suicide – appears to end life only from the myopic viewpoint of an ego on the line. It is actually the scheming of reproduction, a mere moment in the redundancy of the same. Death and birth are indistinguishable, like beginning and end in a perpetual cycle where simple convention discriminates between departure and arrival. Loftiness is debased, the artwork rendered idle, civilisations ruined and sanctuaries abandoned. But bitterly diagnosing the inevitable decline, expending the mind's energy in probing 'the crisis of spirit' is to let oneself off lightly as did Valery's Monsieur Teste. The worst gets missed. And the worst is this: that in the ostensibly mute swamp where everything gets engulfed, larvae stagnate by the billions, fomenting renewal. Plants, animals, and humans, and cultures: everything will begin again. Plots resume. In the reconstructed castles and sculleries Shakespeare's heroes and valets replay their tragedies and comedies, once again for the first and last time. What kind of eye is needed in order to envision a story of such inanity? – The view point – fascinated, horrified – of an idiot. (Lyotard 2002: 12)

This is a noisy cosmological time that is 'ostensibly mute' only in terms of its not saying anything. But in so far as it vibrates it constitutes noise, and as such may not be conducive to observation by the eye, but a listening by the ear as subject and object merge. The intention here is to ensure that 'the worst' does not get missed. On the contrary, 'the worst' is an essential aspect of any dynamic system, and is always present in noise as 'fomenting renewal'.

And so, let us return to Johan Johannsson. He listens to the 'buzzing' of the IBM 1401's world and his encounter challenges the notion of endpoints as the machine finds its voice post-mortem. In doing so it gives rise to a different kind of time that, like Lyotard's temporality, enacts a radical reconceptualization of beginnings and ends.

Johannson's second instalment in his planned trilogy was Fordlandia (2008). This piece charts, like Lyotards rotting pit described earlier, a descent into ruin. Moving backwards in time it identifies themes that can be echoed and amplified as a means of understanding the present while actively pursuing the future. Fordlandia works in multi-temporal dimensions to create a mode of engagement that can reveal without the imposition of universal truth.

> For an album with no lyrics, Jóhann Jóhannsson's *Fordlandia* certainly has a lot of things to say. Narrative grandeur abounds: a Fitzcarraldo-like Brazilian jungle scene, a 19th-century poet mourning a fallen god, an unlucky rocket scientist, the mysterious work of a deformed German physicist. Lumped together, these ideas, expressed in the song titles and exhaustive track-by-track notes which Jóhannsson provides online, sound like the stuff of a Pynchon novel or a master's course, a full conceptual plate that would make the album intimidating if the music wasn't just as beautiful even without its ideas. Jóhannsson's songs are majestic in their simplicity. (http://www.slantmagazine.com/music/review/johann-johannsson-fordlandia)

Once more there are echoes of Jules Romain, and the jungle settlement built on a preconceived sense of reality and the imposition of both an aesthetic and a political economy in Donogoo Tonka. But it did not sound like the jungle said the BBC in its review which said: '[The] album doesn't particularly evoke images of this unusual place and time. If it's steamy, then it's due to hot springs spurting in cold Northern atmospheres, rather than the humid sweating of South American rubber trees. Essentially, this lack of audio imagery isn't so important, as Fordlândia possesses its own distinct charms as a suite-like sequence of pieces' (http://www.bbc.co.uk/music/reviews/jr54/). The lack of 'audio imagery' as it turns out is important in so far as it requires us to engage with the work as a non-representational entanglement of both space and time where Iceland and the rain forest connect across several decades as a number of themes emerge through sound.

The album is in fact multi-themed. Everything is connected to everything and the specifics of the title merge and resonate with other experimental ideas relating to rocketry, the occult and beyond the speed of light travel.

Johannson explicitly engages with the idea of failed utopia, as represented by 'Fordlândia' – a rubber plantation Henry Ford established in the Amazon in the 1920s, as both a means to control the rubber supply for the manufacture of tires and to establish a kind of American idyll as a means of establishing an aestheticized political entity/object that could exist in a kind of vacuum – as an object untouched by and unconnected to other objects. Its ultimate failure, however, was testament to the fact that objects, idealized or otherwise cannot remain impermeable to related elements – elements of nature as well as culture, politics and economics.

Johannsson remarked:

> It reminded me of Werner Herzog's Fitzcarraldo, this doomed attempt at taming the heart of darkness. The remains of the town are still there today. The image of the Amazon forest slowly and surely reclaiming the ruins of Fordlândia is the one that gave spark to this album. For the structure and themes of the album I was influenced by the films of Alejandro Jodorowski, Herzog and Kenneth Anger. I was interested in a kind of poetic juxtaposition and an alchemical fusion of themes and ideas, which I feel is similar to the way Anger uses montage as an alchemical technique – as a way of casting a spell. During the making of the album, I also had in mind the Andre Breton quote about convulsive beauty, which he saw in the image of "an abandoned locomotive overgrown by luxurious vegetation". There is a strong connection to the 'IBM 1401' album in terms of both thematic and musical ideas and I see the two albums as belonging to a series of works. https://fluidblog.wordpress.com/2008/11/10/johann-johannsson-fordlandia/

This alchemical fusion and interconnectivity of creative forces that speak to one another is a feature of noise, as it is developed here. As such Johannson's notion of multimodal temporality stands in stark contrast with Attali's modernist notion of succession. *Fordlandia*, or any of the other music that has been considered here, does not herald, but is rather a complex sounding of infinite referrals. In fact, once you start thinking *through* noise, sound and music, the herald becomes redundant as time becomes non-linear. In a world where noise, sound and music are so diverse, potential heralds multiply to such an extent as to render them virtually meaningless. If music does herald then it has heralded not a new code, but a world that is infinitely complex and unpredictable and noisy. But as fate would have it such a world is understandable via recourse to noise itself, where even if reliable predictions are rare, patterns can be identified and mapped – to a point – in ways that can help us make sense of our times, where *us* denotes an aspect of a pattern within a far-reaching plenum through which everything is present in everything else.

Notes

1 It is a pattern that can link to more fully algorithmic digital accounts that demonstrate the continuous discontinuity of the interface between analogue and digital. http://everynoise.com/engenremap.html

2 D. Nolan, *I Swear I Was There: The Gig That Changed the World* (IMP 2006).

3 https://www.newscientist.com/article/mg21929254-600-the-wonder-year-why-1978-was-the-best-year-ever/story book.

4 This was also the case for Joy Division, of whom Tony Wilson, Head of Factory Records said, referring to their performance at a Manchester battle of the bands contest: 'They had no other choice but to be there'. Theirs was not a contrived attempt to author their own intervention, but a resonating with the times, which was as much a part of them as they were of it.

5 The year 1982 was the official Year of Information Technology in the United Kingdom, so parallel developments in music rather than being regarded as a herald were accompanying activities in the wider political economy.

6 As a development of Witmore's advocacy of sound enhanced archaeological practices and its links to 'post-rock' see, http://www.interferencejournal.com/articles/a-sonic-geography/the-sound-of-ruins. 'Music sounds out space in fundamentally dynamic and often disjunctive ways, it is not simply a product of its environment. A genre such a post-rock attests to this: beyond the "cinematic" aesthetic that is a formal hallmark of the genre, post-rock neither simply reflects or represents pre-existing spaces, but instead stages an elegy for and symbolically reconstructs those times and places lost under late capitalism' (Flectcher 1).

7 The term 'world music' was coined in the1980s to market music as a global phenomenon to a newly curious market. See Chris Nickson, *The NPR Curious Listener's Guide to World Music* (Grand Central Press, 2004), pp 1–2. Veit Erlmann, 'Aesthetics of the Global Imagination: Reflections on World Music in the 1990s', *Public Culture*, vol. 8, no. 3 (1996): 467–488. Simon Frith, 'Born and Hesmondhalgh', in *The Discourse of World Music* (University of California Press, 2000). R. Nidel, *World Music: The Basics*, 2004.

8 http://cdm.link/2011/06/future-shock-the-emergence-of-detroit-techno-told-by-wax-poetics/.

9 http://thequietus.com/articles/18922-oneohtrix-point-never-garden-of-delete-interview.

10 *Donogoo-Tonka or The Miracles of Science: A Cinematographic Tale* is a mock film scenario written by the French novelist Jules Romains. The satirical plot concerns a famous geographer whose academic career is about to be derailed by the revelation that he has invented a city in South America. Through a hilarious sequence of events, a suicidal young man finds a new mission in life by undertaking to find the fictional city, thereby redeeming the error and reputation of the absentminded professor.

11 Contemporary minimalist composer Max Richter has written a new piece entitled *Sleep*, which is eight hours in length and designed to soundtrack a night of slumber. Richter describes the piece as 'an eight-hour personal lullaby for a frenetic world and a manifesto for a slower pace of existence'. The piece is scored for piano, strings, electronics and vocals. It is warm yet haunting and melancholic, while moving at a glacial pace. The German-born British composer said something during our conversation that stuck with me: 'I think time is a preoccupation right now.' This brought a documentary about Marina Abramovic – *The Artist Is Present* – to mind. The film, released in 2012 chronicles a retrospective at MOMA and features a 736-hour static, silent performance piece, which sees Abramovic sitting immobile in the museum's atrium during opening hours while spectators were invited to take turns sitting opposite her. As people stared back at her, some for a few minute, some for hours, many began to cry or describe a feeling of transcendence afterwards. Richter has previously described his music in the terms of storytelling but this time, with *Sleep*, his most ambitious piece to date, it's like he has deliberately left pages blank. It is as if a projector with no film is beaming a hazy light onto a blank cinema screen for you to project your dreams onto. (http://thequietus.com/articles/18458-max-richter-interview).

12 https://genius.com/Max-richter-the-blue-notebooks-lyrics.

13 Zygmunt Bauman also describes how stories serve to compress human knowledge and in the process produce waste. See Zygmunt Bauman, (2004). *Wasted Lives: Modernity and its Outcasts* (Polity, 2004), pp. 17–19.

4

Continuous Discontinuity: A Non-Linear History of Noise

Much of what has been said so far alludes to the question: What do the sounds we hear today tell us about what is to come? Are they naturally orientated towards the future? An engagement with sound as future-orientated or as heraldic draws us into a relationship with music as an ordering of chaos, a calming of noise where what is heard today will be seen tomorrow. When it is eventually seen, it is at the expense of the unseen. Music then is often used as a metaphor for a process of organizing and *filtering out* that which eventually attains a higher existence on a visual plane in line with the thinking of Paul Klee for whom sound, as it emerged from noise, and music were interim staging posts on the way to logical reasoning as picture. Understood in this way, the filtering out or organizing of elements into a coherent pattern is epistemologically privileged at the expense of the shadowy, the discarded and the deemed unnecessary.

Paul Klee regarded music as operating at the vanguard of artistic practice. For him it set the standard for all other practices that would follow and strive to match its achievements. Klee saw music as having reached its creative zenith in the eighteenth century with Mozart. For him contemporary music offered no critical opportunities. Instead it was in painting that such opportunities lay. But only in a particular kind of painting that would draw on the 'golden age' of music that had preceded it. 'In Klee's eyes, Mozart achieved almost superhuman dimensions as his music heralded the fusion of the heavenly with the earthly (or the infernal), an ideal that Klee himself sought to attain as an artist' (Duchting 2004: 8). For him Mozart was the pinnacle of artistic expression. Later music inspired him less. Contemporary music of the nineteenth century, including Bruckner, Wagner, Mahler and Strauss were examples he thought, of a decline in creativity. Like Attali he found modern music too academic, too caught up in theory to serve as a model and source of inspiration in the way that Schoenberg's system of composition had done for Kandinsky. Music at its best could transcend theory he thought.

The overarching aim here has been to stress the importance of music within the more general field of political economy. As such any distinction between entertainment and 'serious' or theoretical music has been avoided for the most part. Despite favouring in my examples music that might be considered alternative, marginal or even intellectual, my intention is not to make any such distinctions but instead to allow all music to resonate in a complex environment. Music that is not on my register, or which I regard as insignificant or inferior (and I admit to making such judgements) is just as significant, maybe even more so in a sonic economy where everything is connected.

In his explorations of sound, Klee specifically drew on musical concepts of rhythm, and perhaps more importantly here, time:

> He referred back to technical terms used during his studies with [Heinrich] Knirr which emphasized the sense of time in the process of painting: the process of creating an image, the expressive strokes of the brush, the genesis of the final effect. Music took on the role of meter by demonstrating the temporal process within a piece of music and also by serving as a metaphor for a new, creative painting that could reveal the temporal aspect with its own unique means. (Duchting 2004: 10)

Klee in his early years knew he wanted to translate the power of music into a powerful new kind of painting, but he was not sure how. This was not to be a grand total artwork of the kind envisaged by Wagner; not a mix of audio and visual qualities; not a way of painting *like* music, or *about* music, or *influenced by* music, but a way of channelling the energy *of* music at its best *through* painting.

Abstract art in so far as it eschews representation has a long affinity with music, and there have been many attempts to imbue artistic practice with the perceived pure spirituality of music. Ludwig Tieck and Wilhem Heinrich Wackenroder notably sought a defined 'poetical' quality that might be applicable to both painting and music – one that would remove the imitative principle and replace it with a renewed sense of freedom (Mininger and Peck 2016). Wackenroder's fictional biography *The Strange Musical Life of the Composer Joseph Berglinger*, completed the year before his death, established most of the key terms for the German romantic conception of music. Their thinking owed much to late eighteenth century notions of the *sublime* as that which surpasses human cognition, and leads to the loss of subjective control and its replacement by feelings of ecstasy. Music from this perspective was regarded as having essential qualities that were present in its orderliness, implying that music conveyed a kind of experience that was irreducible to language. It was thought to be able to transcend worldly concerns. Such qualities are often taken for granted, but the intention here has been to challenge such assumptions. Music, and sound as Cohen (2010) argued, rather than being essential are aspects within an immanent and noisy sonic economy wherein complex and multisensory interactions are processed to form patterns. These patterns are dynamic and constantly renewing, they encourage recognition while simultaneously escaping our grasp.

Klee sought the solution to developing such a comparable organized pattern that he identified in music, for the visual arts, and he sought it in the temporal phenomenon of rhythm. The figure of rhythm was significant in the early twentieth century, largely due to the work of Henri Bergson and Friedrich Nietzsche. Klee noted the ability of the polyphonic fugue to impact on our experience of time, noting that Mozart and Bach were more modern than the nineteenth century. They and their music, he thought, were essentially ahead of their time, and the world needed to catch up. The polyphonic fugue used a number of basic rules, a clear structure and variations on a theme, to fuse temporal flow. Klee developed similar patterns to inform his painting. He aspired to virtuoso status in terms of his control of the palette, using the baroque analogy of the piano keyboard as a simple mastery of his instrument that would afford the purest form of expression. This systematic and structured nature of music – in certain forms – began to influence his own work. He developed a technique based on mathematically proportioned ratios of black and white to inform his black watercolours of 1908–1910 with each shade being attributed a tonal value. This mathematization of his process is akin to contemporary digital technologies such as grayscale and he would spend the years that followed trying to integrate colour into this system of order and value.

In pursuance of his system Klee worked closely with Kandinsky, who also sought a formal theory of sound based on colours. Drawing heavily on Schoenberg's twelve-tone system his aim was to purify art of its 'earthly deficiencies'. For him painting was regarded as a 'thunderous collision of differing worlds; a clash whose outcome is the creation of a new world which we call a work of art. Technically, each piece comes to being just like the earth was made – from catastrophes which can produce, out of the cacophony of instruments, a symphony, which is called the music of the spheres' (Duchting 2004: 20). This was a dialectical approach through which clashing components resolve to create something new. But as the discussion of dialectics that was had earlier has shown, they do not always get resolved. They remain like the unresolved temporal fissures that Lyotard proposed, as part of a noisy environment where rhythm, movement and vibration form patterns or symphonies without conventional recourse to cause and effect or temporal linearity.

For Kandinsky and those who inspired him, the adoption of dialectical thinking could generate a total system of art, and in turn a reconstituted political system. The systematic coding of musical qualities could serve as a model for future developments in matters relating to political economy in much the same way as Attali described. Quoting Charles Péguy (1910), Alex Ross says: 'Everything begins in mystique and ends in politics' (Ross 2012: 386). So music, once thought of as transcendent and spiritual becomes immanent and political. Ross describes how at the turn of the twentieth century wave after wave, beginning with Schoenberg and Stravinsky, of mystical and experimental music came to be politicized. 'All the informational clutter of late-capitalist society, from purest noise to purest silence, from combinatorial set theory to bebop jazz, came rushing in, as if no barrier remained between art and reality' (Ross 2012: 387).

Indeed, there are no barriers between art and reality. Music was probing new possibilities and as Attali has said, offering up new versions of the future, even if, in its theoretical and experimental mode, it was destined to fail. As such it became recognized as significant by political elites, and as an important weapon in the cold war. Whereas, Romanticism flourished in the domain of the totalitarian state, 'free thinking' and experimentalism was developing in the West, and often with state support (Ross 2012: 387). Schoenberg and Adorno, Messaien, Boulez, Cage all pioneered new, avant garde and experimental forms of *future music* that were intended to disrupt the conformity they identified in conventional music. In reality however they were easily absorbed by the very system they were trying to disrupt.

Klee was less interested in such dialectical posturing and instead sought to develop and experiment with ideas relating to musical notation and the divisibility of rhythm as a series of repeatable units to form a method for his art.[1]

> Using Bach as an example, Klee explained the difference between 'individual' and 'structural', or what he termed dividual components of musical composition, concepts that he then applied to the field of painting. The 'structural' or 'dividual' element is part of a larger unit characterised by rhythmic repetitions without variation and hence divisible – it can be divided up into smaller units of the same fundamental structure. The 'individual' components, on the other hand, are defined as a superior, rhythmically independent, unrepeatable an irregular unit of composition, which remains essentially indivisible and can be easily combined with a structural rhythm. (Duchting 2004: 35)

Klee's rhythmic system, drawing on his earlier experiments with black watercolours was an attempt to use the totality of colour as a means of measuring time, as a continuous vibrating flow, punctuated by individual elements. If music was the systematic organization of noise then Klee's art sought a similar status, in terms of being able to organize and express spatio-temporal dimensions not available to representation alone. To achieve his aim, Klee explicitly sought to collapse a very specific dualism. Using Mozart as his guide he noted, 'All that is devilish must be fused simultaneously with all that is celestial . . . Dualism should not be treated as such but rather as a complementary whole . . . Truth requires the consideration of all elements, the work of art a fusion of them all' (Duchting 2004: 52).

This 'fusion of them all' is taken here to constitute noise. The devilish and the celestial, like the audible and the visual, do not reside at either ends of an axis where one prevails in ratio to the other, but constitute a new form of expression, an expression of simultaneity – an expression of noise, or expression as noise.

The heraldic nature of music is once more significant here. It marks a level of achievement that visual art must strive to attain. But if music is a herald or a clarion call for other art forms to emulate then music in itself is regarded as inadequate at some level. It points to what must/can be done: Where 'to show' is the ultimate achievement – to become manifest is the real goal. But it was not a promise that music itself could fulfill. Instead it lapsed into predictable formula for Klee. And this heraldic promise before an eventual lapse is a pattern that endures. Or does it? If this is indeed the case, then maybe that is its strength, its essential characteristic – the offering that is never fulfilled, the end that is never reached. With this comes echoes of Lyotard's critique of incessant and successive endings and beginnings. The end that is never reached, as death, is the temporality of noise.

Klee did not support the notion of a unified total art that combined music and painting, because he believed music and painting were too preoccupied by their own essence to merge. For him the relationship was not prefaced on the notion of similitude – painting was not *like* music. Rather the production of music of a certain kind had an essential quality that he thought could and must manifest itself in other media. In this case, painting.[2] That quality meant that it did not rely on dialectical logic, nor did it require the imposition of a rational representational code to validate it as a form of expression. So, for Klee music was a herald to a new mode of artistic practice in so far as it was able to demonstrate certain qualities that were not yet available to painting.

This interplay between the audible and the visual has been described by Jonathan Sterne as an audiovisual litany. He sets out eleven differences between the audible and the visual, and an analysis of them in relation to the aims of this book will be worthwhile. They are:

Hearing is spherical, vision is directional. This equates to Marshall McLuhan's (1988) 180-degree or 360-degree distinction, and it also links to the originary moment of critical distance as described by Wilfrid Sellars. As part of his 'refinement' phase the chaotic noise inflected theatre of engagement, eventually begins to settle as specific patterns form – patterns that both involve and inform human subjectivity as the gap that affords perspective begins to open.

Hearing immerses its subject, vision offers a perspective. To be immersed is to be surrounded by information in an almost fearful exposed scenario. Like Deleuze's account of the refrain (1987), it is a noisy state of affairs, which is quietened by recourse to organized patterns. The ultimate withdrawal from such a noisy environment is achieved by establishing critical distance and perspective – a safety zone where each individual is placed in direct relationship to the world and where the subject becomes the author of his or her own perspective.

Sounds come to us, but vision travels to its object. The idea of looking upon from a particular perspective contrasts with sound, which attacks from any angle at any time where to meet is also to be met – a situation where the subject is always open (you cannot close your ears). Vision on the other hand can be controlled and directed, it is assumed, as the individual turns his or her head to gaze upon.

Hearing is concerned with interiors, vision is concerned with surfaces. What comes to populate ones' inner world as sounds penetrate involuntarily is a condition of being in the world, being at the mercy of external events, or if not at their mercy as a node in the constantly interlinking of connections that come and go and of which the internal workings of mind try to make sense. With vision the constant uncertainty of pattern formation, reformation as information, is organized in a timely and reliable manner across a 180-degree plane that is constituted in contemporary culture as a flat surface or screen.

Hearing involves physical contact with the outside world, vision requires distance from it. In terms of vibration, sound is always physical and always analogue; it crosses the critical divide and resonates as affect. Humans are in noise, mind and body not split asunder in Cartesian term, but pulsating particles as they are for Sellars and Whitehead. Vision, as has already been determined, withdraws and operates at a distance, where rationality and the thinking subject operate as distinct from the physical reception of light data.

Hearing places us inside an event, seeing gives us a perspective on the event. This being inside is significant as it challenges the Platonic logic of the abstract beyond – beyond the cave to a world outside where vision and logic prevail until the sun becomes so bright as to deter man from further investigation and eventually to the realization that the really real can never be accessed, and thus must be rendered an abstract and theoretical realm. The real on the other hand can be measured and named in an attempt to categorize all that is to be 'known'.

Hearing tends towards subjectivity, vision tends towards objectivity. This aspect of the litany requires that we return to some degree to Sellars and the discussion earlier that outlined philosophy's attempts to reconcile the subject–object relationship. In many ways this is the key aspect of this book as it has been for so many others. This is the reason why noise is invoked as a figure that speaks to the vibrating process of particle formation that defines and supports everything that is and everything that isn't yet. It infers a complex array of uncertain possibilities out of which realities, multiple or otherwise emerge, without deferring to the supremacy of the human as Dasein (Being), but by placing the human *in* noise, *from* noise, as degrees *of* noise in constantly shifting ratio.

Hearing brings us into the living world, sight moves us towards atrophy and death. The constant movement that creates noise and sound (as something perceived) requires constant interaction and engagement. There is no choice but to hear. Once visualized, named, grouped, and categorized however, then the work is done; always to be represented as a code, language and system that requires common contract–consent and not interaction. Such consent implies withdrawal and disengagement, which in turn brings decay. This is significant in terms of political economy and modernism as an emblem of progress that begins to crumble at journeys end (Lyotard 2001).

Hearing is about affect, vision is about intellect. Once more the audiovisual litany seems to reside at the heart of Sellars's attempt to bridge the philosophical gap between subject and object and between the manifest and scientific image. If hearing is about affect then it also resides in the anti-intellectual or Dionysian realm that contrasts to the Apollonian in Nietzsche's *The Birth of Tragedy* (1967). This aspect of the litany draws towards Merleau-Ponty's notion of embodiment, where the body is an instrument of the mind.

Hearing is a primarily temporal sense, vision is a primarily spatial sense. Yet again an aspect of the audiovisual litany is of central concern here. Is sound, hearing and for our purposes noise, essentially temporal, as distinct from the spatiality of the visual? Is music a temporal art in distinction to the visual arts? This has been explored via a discussion of Paul Klee and should also be read within the context of Jonathan Cohen's earlier argument that sound is not uniquely temporal.

Hearing is a sense that immerses us in the world, vision is a sense that removes us from it (Sterne 2003: 15). Questions arising around the impact of immersive media as opposed to more traditional forms are currently of great interest. They form a key aspect of contemporary media discourse as it relates to multimedia practice in general and to visual music practice in particular (Brougher and Mattis 2005). The intention here has been to surmount some of the challenges that are present in this dualism by listening and looking simultaneously through noise.

Sterne goes some way to resolve the dialectic between the audible and the visual, as he resets the ratio of their interaction. But through this resolution he reasserts the phenomenological privileging of the perceiving subject as sound, now returned to its correct historical place, prevails over noise, and so the dialectical process resumes, as the now multisensory rational subject is strategically positioned to meet the next challenge. The aim here though has never been to resolve this dialectical conundrum, never to resolve the problem of noise, but to have it replace dialectics altogether.

Much of what has been discussed so far, even where attempts to collapse dualisms have been convincing though, have accepted temporal linearity and the dialectical nature of resolution as the motor of history. With this comes a certain kind of logic that is determined to bring the universe into some kind of order, to fix it as a knowable space that proceeds through time towards definable and predictable ends. Such logic imposes a critical distance that is essentially visual in nature, and even when it draws on the non-representational qualities of the audible it does so in a way that creates a linear temporality with the audible operating as a herald – a sign (if that sign can be properly decoded) of things to come. This precession of the audible as a feature of Klee's thought, as well as Attali's is also present in contemporary thinking. Returning to Jonathan Sterne, he points out in his *MP3: The Meaning of a Format (Sign, Storage, Transmission)* that although it is now an established technology, the MP3 offers an important point of entry into the interconnected histories of sound and communication in the twentieth century. To access the format's historical meaning, he says, 'we need to construct a new genealogy for contemporary digital media culture. Many of the changes that critics mark as particularly salient aspects of contemporary digital or "new" media happened in audio before they surfaced in visual media' (Sterne 2012). And in a similar vein, Frances Dyson proposes that digital media encapsulate 'an accumulation of the auditive technologies of the past'. For her 'The historical resonance of audio can be extended across the various registers of new media, from their sensual dimensions in both the auditory and visual domains, to their treatment of subjects, to their technical structure and industrial form' (Sterne 2012: 12).

What is evident in these statements is the consistent notion that sound and sound-based technologies precede. So, we find ourselves again embroiled in a temporal chain of events, this time driven by sound. It is still largely understood as a dialectical chain, where the audible precedes but the visual still dominates, or at least strives to dominate. As such the audible and the visual remain distinct and this distinction confounds the possibility of a multisensory mode of engagement with the world.

The figure of noise offers away of collapsing such dualisms into a more complex dynamic system, where new opportunities for critical engagement emerge. Thinking through noise, and in turn through sound and music allows very specific challenges to be made to dominant representational modes of thought. To do so has echoes of the challenge issued by Roland Barthes in his chapter 'The Grain of the Voice' from *Image Music Text* (1993). For Barthes the best way to rearticulate music and its relationship to rational thought was not to reform the language–music interface but to reconstitute music itself with reference to what he called the grain, where the voice is not something used to describe music, but to produce it. Music did not submit easily to the description he thought, and by situating the voice *in* music and in noise rather than as a counterpoint to it, Barthes undermined the dialectical rationale that sought to render music static and generic. Its ordering through language and meaning constitute the movement towards the manifest image and the eventual prioritizing of the visual and the human through the lens of phenomenology. If Sellars went some way to overcome this by reinstating the significance of the scientific, Barthes also contributes by returning the voice to a noisy temporal realm where subject and object need not be separated.

It is dialectics as ordering and resolution that silences or subdues noise, bringing it into the world of the knowable as it is constituted by the interplay of subject and object mediated through language and meaning. It is a world of objects, but also a world where the character of those objects must necessarily be influenced by their being known. And even when modernism gives way to postmodernism, or structuralism to post-structuralism, when the emphasis moves to intertextual combinatorics that create a semiotic surface that can be read and reread, it does so in a way that ultimately reserves the privilege of humans to search for truth and meaning through persistent and disciplined questioning. Such a logical ordering always seeks to compress, remove the noise from what is understood to be an essential signal. To begin to rearticulate noise in the manner that has been argued for here, is then to begin to understand the world that comes after postmodernism. This is important because even in the postmodern moment and in the years since it came to occupy the intellectual imagination, subjects and objects have continued to depend on each other in a relationship that Quentin Meillassoux calls correlationism:

> Correlationism is defined by Meillassoux as the doctrine according to which "we never grasp an object 'in itself', in isolation from its relation to the subject." Kant's transcendental idealism is correlationist, and so is Husserl's noetic-noematic structure. For correlationism, a mind-independent reality cannot exist, because the very fact that we are thinking of such a reality means that it is not mind-independent after all. From the correlationist point of view, Meillassoux says, "thought cannot get outside itself in order to compare the world as it is 'in itself' to the world as it is 'for us', and thereby

> distinguish what is a function of our relation to the world from what belongs to the world alone. (http://www.shaviro.com/Blog/?p=1012)

Correlationism is the quiet coming together of elements that ultimately remain distinct, of music and adjective, story and narrator, contained within thought. And the inability to get outside of thought, to get access to a time before humans prevailed, or to a time in the future where they may no longer prevail, is due to the inability to resolve the bifurcation of the world 'as it is' and the world 'as it is to us' without recourse to correlationism. This is because acknowledgement of a situation, even one that excludes human subjects always puts the perceiving subject who does the acknowledging back in the picture. But what if there were no picture, no fixed representation as image, but instead a series of infinite referrals through noise? This would suppose that noise could exist as pure reason (as Leibniz might have it), untainted by language and representation, in a world without thought, or a world where thought was a passing aspect of noise – a pattern formed temporarily. It would be a world where time was non-linear (DeLanda 2005) and where human intervention, intentionality and control would be minimal at best, and more likely incidental and coincidental. The proposal of such a world does not go far enough for Meillasoux and Harman however. Their real world is a mathematical realm of disconnected objects that do not require human validation on any level. In fact, human subjects are for them an obstacle to knowledge. And this argument is not without merit. Ultimately however humans, whether subject or object, connected or disconnected, do continue to figure largely, even if it is only as events in cauldron of constant renewal as Whitehead believed.

It is not the intention of this book to resolve the debate around correlationism, but rather to propose an approach where the bifurcation that requires resolution via recourse to correlationism in the first place, is replaced by a singular noisy cosmos (Shaviro 2014).

As a way to think further about the interaction between the subject and the object world, and as a means of supporting claims pertaining to a cosmological singularity and to noise, it is worth pausing to consider Jean Luc Nancy's comparing of two passages on the concepts of 'meaning' and 'sound'. He says,

> Meaning: Meaning [le sens] consists in a reference [renvoi]. In fact it is made of a totality of referrals: from a sign to a thing, from a state of things to a quality, from a subject to another subject or to itself, all simultaneously. (Nancy 2007: 7) Sound: Sound is also made of referrals . . . it resounds, that is, it re-emits itself while still actually 'sounding', which is already 're-sounding' since that's nothing else but referring back to itself. (Nancy, 2007: 7–8)

Both meaning and sound then, are for Nancy made up of a series of referrals without beginning or end, in an arena of constant renewal. For Nancy, as for Barthes, this applies to the actual physics of noise and sound as much as it does to the metaphysics of infinite referral as meaning.

As Brian Kane points out,

> Meaning and sound share the 'form, structure, or movement' of resonance. In the perpetual movement of this meaning, Nancy avoids the adequation and static presentation that characterize signification, for there is no closure in the economy of resonance and renvoi. If the phenomenological subject got what it deserved – a static sonorous object – then mutatis mutandis, the same follows for the resonant subject. Nancy conceives the subject, not as a proper self (an I), but as a 'form, structure, and movement of an infinite referral [renvoi], since it refers to something (itself) that is nothing outside of the referral'. (Kane 2012: 446)

The very idea of the self is questioned by Nancy, as it was by Lyorard, in a way where to position oneself as narrator, able to represent the self, is always problematic for a subject that cannot hear itself, but is heard by and can hear only others. Hence, it is in the dynamic relations or referrals that the subject becomes anything meaningful at all as an instance of what Kane calls an 'ongoing temporal or rhythmic pulse'. This rhythmic pulse that resonates with other pulses, human and non-human means: ' . . . sound is not "intentioned" by the subject, rather it is contemporaneous with the subject because meaning, sound and self all share the same "form, structure or movement," namely renvoi, resonance' (Kane 2012). This is interpreted by some as meaning that we live by making music out of a noisy world.[3] But for Nancy, and for Lyotard, as it had been for Whitehead, the repositioning of the subject within a noisy vibrating cosmos goes beyond resetting the phenomenological clock to include the audible. We do not make music in a noisy world, we are music in a noisy world.

Referring to Nancy's critique of Pierre Schaeffer in this regard, Kane has said that although there are similarities in their approaches, there are also subtle distinctions. In terms of the language deployed, Nancy chooses to use écouter instead of the verb, entendre. Listening, as entendre implied a degree of understanding Nancy states, and with it the presence of meaning. As such it retains the notion of a consciously intending subject with the capacity for attention and the decoding of signification. This in turn means that the subject–object dualism remains firmly intact.

He goes on to say:

> Nancy reads the creation of signification, the adequacy of the sensible and the intelligible, as a will-to-truth. The agent of this will-to-truth is none other than the subject, 'capable of presenting the concept and the intuition together, that is, the one through the other' (Nancy, 1997, p. 23). The close interconnection between the subject and signification – the subject acting as ground which establishes the adequacy of the sensible and the intelligible – also means that critical questions about signification inculcate the subject as well. Nancy is arguing that, in the face of a critique of signification, there is a recoil or decision made by the subject – one that is motivated not only by the desire to suture together the sensible and intelligible as adequate but also by the desire to hold onto the subject as the ground of such adequacy. The subject willfully imposes the closed system of signification in order to reassert and reassure itself. (Kane 2012: 443–4)

Hence, the correlation between the intending subject and the sound object as phenomenological referral should be replaced by a dynamic relationship of movement and encounter where the self is reduced to an aspect of vibrating matter. It is not eradicated entirely but repositioned as an event among events where the subject and object exist simultaneously and on many levels of unresolvable interaction and renewal.

What is at stake here is the extent to which this claim impacts on an assessment of the turn towards sonic thinking as a means of accounting for the all too often assumed discontinuous shift in relation to the *digital turn*, and the impact it has had on human subjectivity. The idea of a continuous, always renewing trajectory with subject and object always entwined as a singular unfolding through noise may be incongruous with the temporal discontinuity of the digital realm wherein man is self-coded out of existence. In his *Adventures of Ideas* (1948), Alfred North Whitehead sets out a model of renewal and change that is particularly relevant in this regard. Referring to the gradual and incessant undermining of [Newtonian universalism], he says,

> It (Newtonian Universalism) has enabled men to obtain a new command over nature. Where we formerly obeyed, we now direct. But at last the Newtonian cosmology has broken down.
>
> The story of the breakdown extends over more than a century. For by far the greater part of that period men of science were quite unaware that the ideas which they were introducing, slowly, one after the other, were finally to accumulate into a body of thought inconsistent with Newtonian ideas dominating their thoughts and shaping their modes of expression. The story commences with the wave-theory of light and ends with the wave-theory of matter. It finally leaves us with the philosophic question, what are the concrete facts which exhibit this mathematical attribute of wave vibration? (Whitehead 1948: 185)

This describes the presence of a wave or particle duality that resonates perfectly with contemporary debates, and with the aims of this book. The digital turn and related scientific developments throw up a number of very significant issues here. Not least the position of the human subject in a post-human universe, but also the implication of a discrete and mathematical bit mapped model for an understanding of temporality more generally. But when thought through noise it is not a question of continuous or discontinuous, but of a constantly renewing pattern formation. To explain this further requires that we return once more to the thoughts of Henri Bergson and Gaston Bachelard.

Bergson's version of dialectical progression as a continuity, as briefly discussed in Chapter 1, was vehemently opposed by August Talheimer who said: "The fundamental error in Bergson's conception of dialectics is his disregard of the fact that the new which has developed from the old, stands not only in opposition to the old, is not only its negation, but has, at the same time, something in common with the old." And this raises the issue of the continuous or discontinuous nature of change. If the new contains aspects of the old then by virtue of the fact that the constituent parts can be identified and divided, it is discontinuous. The question here in relation to this might be, has the digital replaced the predigital to such a degree that it has destroyed everything we previously thought, or has it retained distinct elements of it?

Talheimer continues:

> If one follows the thought of Bergson, it becomes evident that it cancels itself. There is only one kind of negation in which the thing negated has nothing more to do with that from which the development proceeded. This is complete or unconditioned negation or destruction. If I completely negate a thing, I destroy it, and development is completely stopped. If development is forced beyond its limits, as it is with Bergson, if it is made absolute, it is transformed into its opposite, into fixity or lack of development.
>
> The Bergsonian distortion of dialectics, it is obvious, is very closely connected with the present historical position of the bourgeoisie. The mystical or falsified dialectics of the Bergsonian type rejects historical regularity and replaces it by miracle, arbitrariness, and incomprehensibility whereby nothing is impossible. (https://www.marxists.org/archive/thalheimer/works/diamat/11.htm)

'Miracle, arbitrariness, and incomprehensibility whereby nothing is impossible', is a state of affairs that is of great interest here (miracles aside), as indeed they were for Bergson.[4] Given his interest in time and creativity, although, it is thought provoking to have him described as a destroyer of progress. His is a different kind of progress where newness is constantly regenerating. To say it has no trace of its predecessor may be misleading. In the same way noise operates as the constant, vibrating interaction of elements, the new for Bergson contains elements of the old but without it always being perceptible. Old and new coexist in harmonic resonance. Bergson's destruction then is not a destruction of movement or progress. Instead it is a destruction of beginnings and ends – it is noise.

But others, most notably Bachelard, also questioned how the creative process played out in Bergson's model.

> For Bergsonism the creative value of becoming is limited by the very fact of fundamental continuity. Time has to be left to take its course if it is to do its work. In particular, the present can do nothing. Since the present carries out

> the past just as a pupil carries out a task the teacher has set, the present can create nothing. It cannot add being to being. Here again, Bergsonism has developed by following the intuition of fullness. For this school of thinkers, the dialectic always goes directly from being to being without nothingness being brought in between them . . . We know that for Bergson the idea of nothingness is in the end richer than being [nothingness is occupied by beings that cannot be understood as discrete, but as always in relation each to the other, vibrating and attracting attention] . . . Thus with regard to the knowledge we have of them, no substance could have any void and no melody be broken by an absolute silence. (Bachelard 2000: 17)

This absence of absolute silence is noise as a universal and virtual realm of possibility. But Bachelard thought otherwise. He was committed to reinterpreting Bergson in light of developments in quantum thinking, and as such his thoughts are particularly useful here in light of more recent developments in this area. His was a model of discontinuity, and a universe comprised of discrete elements. As such his thinking, rather than Bergson's seems particularly apt for the digital age. Yet choosing between Bergson and Bachelard may not be necessary.

Teresa Castelaoc-Lawless argues for a discontinuous Bergsonism that moves in and out of dialectics freely and without compromise (2010). In some respects this is not a strategy that Bergson himself would have questioned. Indeed, nor would Bachelard. What emerges then is the possibility, or indeed the necessity for a continuous discontinuity that can account for the complex nature of reality and the role of humans within it. Castello-Lawless states that:

> Bachelard quickly detected the need to accommodate the vocabulary of philosophy to the new sciences in light of their new requirement that scientific concepts be as flexible and mobile as the dialectical mind which produces them, and which demonstrate at different times and in different situations an alternation between rationalism and empiricism, continuity and discontinuity. (Castelaoc-Lawless 2010: 26)

Bergson on the other hand had constructed a dualism between intuition and intelligence, as a double form of consciousness that corresponded to what he understood as the double form of the real. Intelligence, he thought, grasped 'artificial' reality analytically and scientifically whereas intuition grasped reality psychologically and metaphysically. And this equates directly to the thinking of Wilfrid Sellars for whom two such channels merged to form a stereoscopic view.

Dialectics for Bergson though – as the merging of the continuous and discontinuous – represented an unsatisfactory method of analysis. It worked by capturing only fragments, as in a 'still movie', and this was not able to account for a reality and a mind that were in constant flux. Scientific and intellectual investigation, he thought, could reveal only a certain kind of information: Fragmented and static in nature, as it moved through a dialectic in a very mechanical way. Intuition on the other hand, revealed reality on a different register. For Bergson, metaphysics could mediate access to a reality in a way that tended to escape rational categorization. So, for science and the intellect to be of any use he thought, they needed to embrace the qualities of intuition: He believed that 'we needed to keep our intellectual habits from creating images of duration, such as imagining time as moving in a linear spatial trajectory as it was described abstractly in science, instead of the real, concrete time given by intuitive consciousness' (Castello-Lawless 2010: 28). But Bergson also acknowledged that this would not be an easy task. It would involve an intuitive use of scientific concepts. In short, the use of scientific methods in a 'non-scientific' way or in a 'superscientific' way through 'the grain of voice', in which concepts were liberated as mobile and fluid representations.

Castelaoc-Lawless sets out to demonstrate the confluences in the thought of Bergson and Bachelard, and to challenge the often-made assumption that their ideas stand opposed to one another. Given that both were committed to a flexibility of concepts in terms of their ability to respond to states of flux … 'they both would have to agree that either continuity or discontinuity were incomplete conceptualizations of reality. They would also have to agree that sometimes there was unity, sometimes disunity, and sometimes multiplicity in the knowledge of nature and of the self' (Castello-Lawless 2010: 29). The problem in differentiating among their two approaches though, as Castelaoc-Lawless states, was that Bachelard was in large parts an advocate of Bergsonism, aside from the fact that he did not agree that discontinuity could be reduced to an intellectual tool for the capturing of something that was in reality continuous. Bachelard was keen to show that discontinuity corresponded to the reality of the world as it was beginning to be understood by quantum mechanics. He wanted to imbue duration with arithmetic qualities, to make it more accurate and reliable, more predictable and scientific, and this seems apt in relation to more recent attempts to arithmetize philosophical enquiry.[5]

> Bachelard saw our inability to detect position and momentum simultaneously as definite proof that nature was discontinuous, while De Broglie believed that our inability to measure it except with the help from theories of quantum discontinuity was a sign that nature was fleeting, continuous, and unpredictable. Wave mechanics showed that physical entities were constantly 'in progress.' Therefore, De Broglie continued, 'if Bergson could have studied the quantum theories in detail, he would have noted certainly with joy that, in the image that they offer us of the evolution of the physical world, they show us nature in all its occasions hesitating between several possibilities, and he would have undoubtedly repeated, as in La Pensée et le Mouvant, that "time is that very hesitation or it is nothing at all"' (De Broglie1941, 252–253). The very need for wave mechanics to renounce individualization of particles, as well as its inability to follow the evolution of particles throughout time meant that reality was fluid, just like Bergson had suggested (De Broglie 1941, 255). So, it is possible that Bachelard did not give up on Bergson precisely because De Broglie demonstrated that quantum mechanics verified continuity in nature, and thus that Bergsonism might be right after all. (Castello-Lawless 2010: 34)

More contemporary science is less certain that Bergson was right and has demonstrated the existence of both wave and particle qualities simultaneously. Noise, as it is conceived of here, is in part invoked in relation to such contemporary thinking. It is time as both hesitation and as infinite vibrating potential. The idea of hesitation is particularly interesting in so far as it denotes a discrete discontinuous object, moment or turning point without negating the fluid and continuous features of movement and vibration. If an object hesitates just long enough, it might be able to be distinguished from those other objects that it may or may not resonate with. Noise is offered then as a means of circumventing the continuous/discontinuous, wave/particle dualisms. Noise is vibration of matter at the quantum scale where both wave and particle characteristics are present, and it denotes a reality that pervades the scientific and cultural realms but does not rely on either for its validation.

As intuition, it is continuous and only becomes intellectual, dialectical and discontinuous at the point at which hesitation is resolved and a new course/direction is decided on.

This strange situation can be further thought through the concept of quantum noise.

> Maybe quantum noise, like the noise we encounter in daily life, has a meaning that escapes us. It may seem indeterministic, but could be produced by deterministic processes that, for whatever reason, we can't see. It might, for example, be a consequence of living in one of countless parallel universes and not being able to tell which is ours. The noise, in essence, tells us where we live. All those little upticks and swerves in particle behavior are the quirks that differentiate our universe from others, and they are 'noise' inasmuch as our location is pure happenstance, like being put in hotel room 314 rather than 159. The other basic approach is that quantum noise really is meaningless and quantum theory is as indeterministic as Bohr took it to be, in which case the challenge is simply to tidy up the ill-defined concept of observation. In 1986, three physicists – GianCarlo Ghirardi, Alberto Rimini, and Tullio Weber -proposed that not only is quantum noise meaningless, but experimenters don't trigger it. In fact, nothing does. It shows up, completely unprompted – perhaps once every 100 million years for an individual particle. (http://nautil.us/issue/38/noise/the-noise-at-the-bottom-of-the-universe)

Noise then is either a defining feature of a localized realm of infinite possibility or an infrequent disturbance in an otherwise discernible universe. Either way, noise opens itself up to the possibility of both scientific validation (whether meaningful or meaningless, it does exist, and it exists in a way that is independent of human perception) and mathematical modelling. With varying degrees of reliability such definitions of noise bring with them the promise of predicting the unpredictable, as explored by Nate Silver in his book *Signal and Noise* (2012). In this work Silver raises the possibility of using statistics and mathematics to code, model and ultimately predict, a range of events from baseball to elections. The interesting thing about Silver's approach is his emphasis on patterns and the role that humans play in both creating them and recognizing them, or perhaps, more crucially, misrecognizing them. Paradoxically, the art of prediction, like the occurrence of quantum noise, is for him strangely unpredictable. It is chaotic. Yet like in Chaos Theory, patterns do emerge from what Silver refers to as noise, and they do so as signal. The important thing for him is to identify not only signals themselves, but the right signals that can usefully be mined as information and reliable predictors. This remains an important point here. What data can be identified as useful, and what is superfluous, designated as noise? The intention of this book has been to shift the emphasis in this regard, towards the question of what can be understood from the noise itself, and not only from its organization and resolution into code or signal.

Working through the figure of noise has not been a mechanism for developing models of prediction. It has rather been a way of arguing for a multisensual form of engagement with contemporary social, political, economic, technological and aesthetic questions regarding our contemporary existence. Thus, it has been about temporality as defined by movement, vibration, and uncertainty rather than predictability.

In 'A Chronic Condition: Noise and Time', Paul Hegarty brings this notion of temporality directly into an account of Noise. He writes:

> Noise has often been dealt with in terms of its effect on the body, or on parts of it. This means that noise is generally treated as a spatial problem or proposition. As noise is not autonomous but occurs through being perceived, defined, legislated for and against, as noise, this prominence of the physical encounter with noise has led to deep phenomenological insights about its working, but the embodied is not just 'there' in space, it is also 'there' in time. (Halligan and Hegarty 2012: 15)

In this volume noise, as opposed to sound, *is* autonomous. It is not dependent on perception, but rather exists outside of perception. At the quantum level, noise cannot be triggered by experimenters, and is entirely outside of their control. It constitutes then a widening of the field that Hegarty sets out. It is a widening beyond perception. For as Sterne points out, noise is not any more something to be calmed and brought to order because it can merely be made imperceptible: 'Attali's treatise posited the control of noise as a metaphor for all social control precisely at the moment when communication engineers began to articulate a paradigm where noise no longer needed to be eliminated or reduced if it could simply be rendered imperceptible to the ear' (Sterne 2012: 124). What Sterne does not consider however, is the fact that you can turn off the perception, like Maurakami's scientist in *Hard Boiled Wonderland and the End of the World*, but not the noise itself.

In Hegarty's excellent account, noise approaches (but never reaches) and is akin to the impossible in Bataille, Derrida and Nietzsche. It is an always becoming that continues to disturb. Noise as lived and perceived cannot be allowed to plateau, as in Bergson and Deleuze, and still be noise, he says. It can never be calmed and continue to exist as noise. For Hegarty then, the temporality of noise is a kind of being-at-the-edge, an edge that is tantalizingly never reached. But if he is correct, and noise is positioned in its contemporary context as the expression of new forms of existential time as anxious engagement, then what has been the impact of digital noise, noise control, or the conception of the digital *as* noise? Undoubtedly the digital brings new temporalities as multi-temporalities that induce anxiety, but in asking these questions there is a danger of reintroducing the duality of lived existential time (duration) as analogue manifest and continuous, and a measured theoretical impossible time, as discontinuous scientific and digital. What is important here is their irresolvable coexistence as noise – a noise that is both perceptible and imperceptible – not just a *living* at the edge, but a universal edginess that is characterized by both imminent catastrophe and incessant renewal. This can only be comprehended by reinforcing the importance of a continuous discontinuity where the dualities of real and virtual, digital and analogue are surmounted and where the immanence of heterogeneous temporalities are advocated at the expense of a phenomenological verification of one's own being.

If noise is lived as argued by Hegarty, then it is lived through noise itself, as a series of infinite referrals where the subjective listener is a noise-event of vibrating energy as argued by Nancy:

> The subject of the listening or the subject who is listening (but also the one who is 'subject to listening' in the sense that one can be 'subject to' unease, an ailment, or a crisis) is not a phenomenological subject. This means that he is not a philosophical subject, and, finally, he is perhaps no subject at all, except as the place of resonance, of its infinite tension and rebound, the

> amplitude of sonorous deployment and the slightness of its simultaneous redeployment – by which a voice is modulated in which the singular of a cry, a call, or a song vibrates by retreating from it (a 'voice': we have to understand what sounds from a human throat without being language, which emerges from an animal gullet or from any kind of instrument, even from the wind in the branches: the rustling toward which we strain to lend an ear.) (Nancy 2007: 21–22)

Listening, as opposed to hearing/understanding then, is a particular activity that requires the abandoning of subjectivity – a placing of oneself in a vibrating cosmology, to resonate with all else that vibrates. This is being referred to here as noise.

The approach deployed in this book has been to foreground acoustic practices, working through noise, without regarding it as oppositional in relation to visual or other sensual practices. Rather, the concept of noise is rethought within the context of a *sonic economy* (Kennedy 2015) that recognizes the ubiquity of noise as a means of accounting for the current multisensory nature of the digital age and the attendant knowledge economy. A sonic economy is defined here as the organization and disorganization of noise as a chaotic system where patterns are in a constant state of formation. This ever-changing patterning inclines towards a temporal system of movement and vibration (Barthes uses the phrase phonic economy 1993: 183).[6] As an unstable and dynamic state it has clear implications for the status of knowledge in the digital age. Rather than being semi-stable, hesitant, and requiring a paradigm shift to advance knowledge in a linear progression, knowledge is today rather in a constant state of evolution and flux (a condition that prompts engagement with the temporal as an unfolding discontinuous continuity). This is paralleled in the constantly changing digital environment that brings knowledge of the world to us in ways that need careful reassessment. We encounter this world not at a distance, as in a scopic regime, but rather immerse ourselves in a constant flow of information that bombards our every sense simultaneously. Noise eliminates visual perspective and critical distance on which so many previous epistemological and philosophical frameworks have been founded. As both wave and particle, and a specific category within the sonic, it extends out beyond itself, enabling immersive modes not only to be mapped on to existing areas of knowledge, but to simultaneously challenge our ability to make sense of the world according to a taxonomic order which organizes knowledge into discrete units, categories and disciplines.

The position adopted here proposes noise not as something to be eradicated but as something to be potentially utilized as a tool for critical analysis. Noise has to date almost invariably been thought of as the unwanted portion of a signal, as something annoying, undesirable and something to be eradicated. Increasingly, however, across many different disciplines, various forms of noise are now being figured as the most important part of the signal: Neuronal noise, hypersonic silence, dark matter, the black/shadow economy, the detritus left over when sound files are compressed, are all phenomena that are being grouped together here under the heading of noise when this term is expanded beyond its definition as a uniquely acoustic phenomenon. The aim has been to communicate a sense of noise as a cross-cultural, multimedia, multisensory, transdisciplinary, cross-modal phenomenon that problematizes the assumptions made by dominant discourses or disciplines working in isolation and in which anything that does not fit neatly within the pre-established terms of that discourse or discipline are figured as an unwanted irritant to be gotten rid of, as noise.

Noise then becomes the figure through which different forms of interdisciplinary, transdisciplinary work can be enabled, mixed, and remixed. As Arthur Kroker in his essay *Codedrift* says, 'Encoded by technology, everyone today is a code drifter, touched by technology and remixing the technology right back.'[7] Everyone is connected to everything – 'touched by technology', humans exist in a complex nexus with each other and with external objects he says. Objects do not retreat into their own internalized vacuum, as Harman thought, but resonate outwards to form connections, that are dynamic and temporal as Whitehead believed.

To this end it is essential to consolidate noise as praxis and to explore the possibility of reconfiguring it as a means by which social, political, economic, technological, scientific and aesthetic phenomena can be better understood in their contemporary context. To do so is to prevent noise from being reduced to a meaningless cacophony of merely subjective points of view, this being its present fate as argued by Greg Hainge when he writes: 'This, however is just how noise has been treated, as it has been used to apply to everything and nothing at the same time, subject to a whole host of mutually contradictory definitions and usages, its apparently ineffable nature the result of divergent agendas rather than something proper to noise itself' (Hainge 2013: 8).

Hainge goes on to suggest that noise is in fact the unavoidable substrate of all existence. This generalized, rather than particular conception of noise, supports the development of a model based on the complementary idea of noise as a driver, an energy, or as Jane Grant (2013) has suggested, a 'buoyancy' which enables systems to be alive and to adapt and change and interact with their environments. There is something significant about noise that makes it uniquely able to account for the complexity and chaos of the contemporary age. The turn to noise, sound and music as a means of accounting for such complex digital environments is useful because it allows for engagement with the not readily present and invisible realm that is immanent rather than transcendent, as was the case for Heidegger and Derrida, but this time in a way that is less dependent on the phenomenological subject.

Within such a complex nexus, time is both analogue and digital, moving as both wave and particle. Analogue time flows continuously whereas the digital enacts a discontinuous fracturing. But if we think of data as a wave, or a datawave, or as possessing wave/particle qualities then we can account for a condition of existence where analogue and digital move in and out of each other almost seamlessly. Again, Kroker gives an example where sound is a wave but can be deconstructed digitally as if made up of distinct particles and reassembled with startling accuracy. He says, 'Consider this description of a newer technological innovation – hypersonic sound, unidirectional sound: Beaming waves of hypersonic sound at a pitch that is undetectable by the human ear. The waves combine until they smash into an object such as a person's body. The waves then slow, mix and re-create the original audio broadcast. If the person steps out of the waves, they are no longer obstructed, and are rendered inaudible' (Kroker and Kroker 2010).

This is significant in terms of the interaction between humans and their environment, or between humans and objects, and between humans and technology. Many of the things that humans interact with are not perceptible, but they may still be significant in some way. They may also be able to be perceived by machines, and then rendered perceptible for humans.[8] This can be understood as a form of noise that is not constrained by either continuous or discontinuous ties. Noise, sound and music emerge out of the movement between analogue and digital modes, or between the perceptible and the imperceptible as incessant unresolvable interaction and renewal as energy, and not as dialectical opposites. In doing so they create a unique temporality or multi-temporality, that is complex and chaotic, but that is often paradoxically patterned and therefore amenable to modelling that can render it predictable. For some, as has been discussed here, this notion of predictability is most clearly evident in music.

To explore this notion of predictable evolution as essentially a musical phenomenon it is worth returning one final time to Attali's premonitory account of noise and music. It was an account that retained the perspective of the dialectic in relation to political economy. Referring to music Attali said: 'It heralds, for it is prophetic. It has always been in its essence a herald of time to come' (Attali 1977: 4). In saying this Attali followed a broadly dialectical and sequential model. He ended his work on noise by hoping that the conditions brought about by the shift from repetition to composing would be such that a radical reorganization was not only possible, but likely. This dialectical progression has been the primary topic of concern for this book.

In his *Pop Music, Pop Culture* Chris Rojek sets out the dialectical nature of Attali's thought saying: 'Attali is a consistent dialectical structuralist in positing that music is dichotomous. Through melody and harmony, music produces a sense of order and wellbeing; but by means of clamour and dissonance it speaks a different kind of language – one concerned with interference, interruption and rupture' (2011: 60).

Music for Attali assuaged fear as a primal mode of sacrifice and was a model for the measured control of alienation from the means of production and the unpredictability of nature. By invoking a dialectical relationship between culture and nature it assured not just calm but an orderly, linear and predictable process of temporal progression.

> Although he considers music to be the result of the mode of production, Attali claims that changes in music occur far more rapidly than in social, economic, and political infrastructure. Hence, changes in the organization

> of noise and in the nature and technology of music clarify comprehension and prediction in relation to the evolution of a society as a whole. In Attali's work pop is assigned a *prophetic* quality. In this sense 'All You Need is Love' by the Beatles prophesised Gorbachev's policies of *glasnost* and *perestroika* in the Soviet Union two decades later. Music has the capacity to prefigure changes in the concrete social totality. (Rojek 2011: 60)

This is a somewhat simplistic analysis of Attali's stated aims, and perhaps stands as an example of why he did not address music itself more directly in his analysis. Such complex claims are difficult to support in terms of cause and effect because music and other significant events, in their digital and analogue modes, are entangled and non-linear rather than conveniently sequenced. Such entanglement is characteristic of the unique temporality that is created by the fluid movement between noise and music and between analogue and digital that allows the past, present and future to come into relationship with each other in new and interesting ways. Accounting for such fluid movement, and the patterns it creates, can now be accomplished using algorithmic models that may provide opportunities for making manifest theoretical predictions about events before they unfold.[9] And now that music itself has been reduced to a set of key characteristics that can be given a mathematical value, they can potentially be mapped on to other coded values to ascertain what if any, links and relationships can be deduced.[10] If music heralds then, it would seem that we are on the verge of being able to predict the future, as Attali had hoped. But ironically this has come at a moment when the very technological advances that make this possible have also generated a temporal configuration where past, present and future need no longer be neatly arranged as a predictable linear unfolding.

An engagement with this contemporary mode of temporality through noise is prefaced on a reformulation of the present and its relationship to the past and the future as a discontinuous continuity where patterns form and reform in time with a complex, sometimes regular, sometimes irregular, rhythm. It is a temporality of randomness and apophenia, of pattern (in) formation in chaotic environments that exhibit creative energy flows where formation and deformation occur at varied and variable speeds. It is a mode wherein noise is always simultaneously present and absent. And this strange state of affairs, as Kroker describes, can now be specifically revealed and manipulated by technology. As such noise, as a means of grasping both the perceptible and imperceptible, is a useful model for understanding contemporary media environments where fixed, spatialized, and universal patterns of engagement are being superseded by fluid, temporal and relative patterns where the critical distance of the visual realm with its reliance on a rational subject is being replaced by the more immersive characteristics of the sonic and the hypersonic. Engaging with noise in this way thus becomes a technological encounter with temporality rather than a purely existential one. Within such an unstable temporal framework, music, rather than operating as a simulacrum of order and predictability, as Attali thought, is a dynamic resonating *of* time, the art that is made out of time as it emerges from noise. Through its emerging it can reveal, if only fleetingly, significant information about the related conditions of its existence. But as an aspect of noise and not its resolution it is contemporaneous with those conditions rather than prior to them and as such it is an unresolvable mode of immanent engagement.

But what is to be gained by moving to a mode of thinking where nothing is regarded as inevitable or predictable, where everything changes all the time – to a mode of thinking that invokes scientific debates to support notions of relativity, uncertainty and that draws attention away from common sense and instead towards the supersensual, to the world as it exists beyond human comprehension (in hypersonic silence)? Currently most research that deals with contemporary issues of political economy, technology and the dialectical models that support them, do so by reducing or managing complexity through a process of compression. That is to say, vast amounts of information are analysed to identify the most pertinent bits, the bits that can commonly be agreed to be significant. This is mostly measurable, as bits, quantifiable and subsequently amenable to representation, usually in the form of visualizations. The rest of the data – deemed surplus to requirement – is designated as *waste*, or noise if it does not conform to the pattern. Such waste material is often qualitative, complex, fuzzy and difficult to map. But it is precisely this kind of material that is important, because however well it is disposed of, however successfully it is designated as existing outside of the system, on the periphery of the threshold, it always remains and will always resound in noise when noise is conceived as the interaction of all elements within a system rather than an external infringement. Moving noise from the periphery to the centre then, allows political, economic and aesthetic judgements to resonate more directly with our contemporary existence. It is an existence that can be heard in and through noise, sound and music as foundational elements that serve to energize our non-linear temporality in a way that celebrates uncertainty rather than predicting reliable and quantifiable outcomes.

Notes

1 The ideas of *dialectical synthesis* and *rhythm* have been explicitly combined by Henri Lefebvre. Instead of synthesis Lefebvre showed how rhythm could sit between a thesis and an antithesis and in the process modulate or keep time so as to regulate the ebb and flow. Rhythm was a third element in a triadic system (Rhythmanalysis 2010).

2 So again, the dialectical relationship between music and art presents itself. This requires that the notion of synthesis be considered. Browning and Kilmister give an excellent account of synthesis in their *A Critical and Post-Critical Political Economy* (2008: 34). Do music and visual art synthesize to create a new advanced art form? I would suggest not.

3 See http://nautil.us/issue/38/noise/how-noise-makes-music.

4 The question of false problems may be relevant here. Humans state problems that contain the means of their solving, and in doing so a number of assumptions are common – order from disorder being from non-being . . . Bergson takes a different position (Deleuze 1991: 18).

5 Badiou (2005).

6 What is the difference between phonic and sonic? Phonic relates to speech sounds and therefore speaking subjects, whereas sonic is a broader term that does not require subjective presence and is more aligned to wave and vibration.

7 http://ctheory.net/articles.aspx?id=633.

8 As in the case of the sound of the stars . . . Listening machine.

9 https://acloserlisten.com/2017/02/20/where-will-music-head-next/.

10 Digital music: http://www.digitalmusicnews.com/2016/05/17/music-genres-three-attributes/.

Bibliography

Adorno, T. W. (1993). *Hegel: Three Studies*, trans. S. W. Nicholsen. London: MIT Press.

Adorno, T. W. (1981). *Negative Dialectics*. London: Continuum.

Attali, J. (1977). *Noise: The Political Economy of Music*. Minneapolis: University of Minnesota Press.

Bachelard, G. (2000). *The Dialectic of Enlightenment*. London: Rowman & Littlefield.

Badiou, A. (2005). *Being and Event*. London: Continuum.

Barker, T. (2012). *Time and the Digital: Connecting Technology, Aesthetics, and a Process Philosophy of Time*. Hanover: Dartmouth College Press.

Barry, R. (2017). *The Music of the Future*. London: Repeater.

Barthes, R. (1993). *Image Music Text*. London: Fontana Press.

Bauman, Z. (2004). *Wasted Lives: Modernity and its Outcasts*. Cambridge: Polity Press.

Benjamin, W. (1970). *The Author as Producer*, http://www.berk-edu.com/VisualStudies/readingList/06a_benjamin-author%20as%20producer.pdf (accessed 03 January 2018)

Benjamin, W. (2008). *The Work of Art in the Age of Mechanical Reproduction*. London: Penguin.

Bergson, H. (2002). *Key Writings*. London: Continuum.

Brassiere, R. (2007). 'Genre is Obsolete'. *Multitudes* 28. http://www.multitudes.net/Genre-is-Obsolete/

Brougher, K. and Mattis, K. (2005). *Visual Music: Synaesthesia in Art and Music Since 1900*. London: Thames & Hudson.

Browning, G. and Kilmister, A. (2006). *A Critical and Post-Critical Political Economy*. New York: Palgrave Macmillan.

Castelaoc-Lawless, T. (2010). 'Is Discontinuous Bergsonism Possible?' *Agathos An International Review of Humanities and Social Sciences* 1(1): 25–37.

Cimini, A. (2012). 'Vibrating Colors and Silent Bodies. Music, Sound and Silence in Maurice-Merleau-Ponty's Critique of Dualism'. *Contemporary Music Review* 31(5–6): 353–370.

Cohen, J. (2010). *Oxford Studies in Metaphysics*. Vol. 5, http://aardvark.ucsd.edu/perception/sounds.pdf. (accessed 01 November 2017)

Cox, C. (2011). 'Beyond Representation and Signification: Toward a Sonic Materialism'. *Journal of Visual Culture* 10(2): 145–161.

Delanda, M. (2005). *A Thousand Years of Nonlinear History*. New York: Swerve.

Deleuze, G. (1991). *Bergsonism*. New York: Zone Books.

Deleuze, G. (2004). *Difference & Repitition*. London: Continuum.

Deleuze, G. (2006). *The Fold: Leibniz and the Baroque*. London: Continuum.

Deleuze, G. and Guattari, F. (1987). *A Thousand Plateaus: Capitalism and Schizophrenia*, trans. B. Massumi. Minneapolis: University Minnesota Press.

Döbereiner, L. (2014). 'How to Think Sound in Itself? Towards a Materialist Dialectic of Sound', Proceedings of the Electroacoustic Music Studies NetworkConference Electroacoustic Music beyond Performance, Berlin, June 2014.

Duchting, H. (2004). *Paul Klee painting Music*. London: Prestel.

Dyson, F. (2008). *Silent Theory, Technology, Philosophy in Kroker & Kroker's Critical Digital Studies: A Reader.* Toronto: University of Toronto Press.

Dyson, F. (2009). *Sounding New Media: Immersion and Embodiment in the Arts and Culture*. Berkeley: University of California Press.

Epstein, J. (2014). *Sublime Noise*. Baltimore, MD: John Hopkins University Press.

Ernst, W. (2016). *Sonic Time Machines: Explicit Sounds, Sirenic Voices, and Implicit Sonicity.* Amsterdam: Amsterdam University Press.

Foucault, M. (1995). *Discipline and Punish: The Birth of the Prison*, trans. A. Sheridan. New York: Vintage.

Goddard, M. Halligan, B. and Hegarty, P. (2012). *Reverberations: The Philosophy, Aesthetics and Politics of Noise*. London: Bloomsbury.

Grant, J. (2010). 'Foucault and the Logic of Dialectics'. *Journal of Contemporary Political Theory* 9(2): 220–238.

Grant, J. and Matthias, J. (2013). 'Plasticities and Ghosts: Relationships between Stimulus and Memory in Noisy Networks'. *Leonardo Music Journal* 23: 9–10.

Hainge, G. (2013). *Noise Matters: Towards an Ontology of Noise*. London: Bloomsbury.

Harman, G. (2002). *Tool-Being: Heidegger and the Metaphysics of Objects*. Chicago and LaSalle: Open Court.

Harman, G. (2010). *Towards Speculative Realism: Essays and Lectures*. Winchester, UK: Zero Books.

Heidegger, M. (1977). *The Question Concerning Technology and Other Essays.* New York: Harper and Row.

Ihde, D. (2007). *Listening and Voice: Phenomenologies of Sound*. Albany: State University of New York Press.

Ikoniadou, E. (2014). *The Rhythmic Event*. London: MIT Press.

Ingold, T. (2000). *The Perception of the Environment: Essays in Livelihood, Dwelling and Skill*. London: Routledge.

Janus, A. (2011). 'Listening: Jean-Luc Nancy and the "Anti-Ocular" Turn'. *Continental Philosophy and Critical Theory Comparative Literature* 63(2): 182–202.

Kane, B. (2012). 'Jean-Luc Nancy and the Listening Subject'. *Contemporary Music Review* 31(5–6): pp439–447. October–December 2012. Routledge.

Kennedy, S. (2015). *Chaos Media: A Sonic Economy of Digital Space*. New York: Bloomsbury.

Kroker, A. and Kroker. M. (2010). http://www.ctheory.net/articles.aspx?id=633. (accessed November 2017)

Latour, B. (1993). *We Have Never Been Modern*, trans. C. Porter. Cambridge, MA: Harvard University Press.

Le Guin, U. (1974). *The Dispossessed*. London: Gollancz.

Lyotard, J-F. (2002). *Soundproof Room: Malraux's Anti-Aesthetics*. Stanford, CA: University of Stanford Press.

Mauch, M. MacCallum, R. M., Levy, M. and Leroi, A. M. (2015). 'The Evolution of Popular Music: USA 1960–2010'. *Royal Society Open Science*. London. http://rsos.royalsocietypublishing.org/content/2/5/150081

McLuhan, E. and McLuhan, M. (1988). *The Laws of Media: The New Science*. Toronto: University of Toronto Press.

Meillassoux, Q. (2012). *After Finitude: An Essay on the Necessity of Contingency*. London: Bloomsbury.

Mininger, J. D. and Peck, J. M. (2016). *German Aesthetics: Fundamental Concepts from Baumgarten to Adorno*. New York: Bloomsbury Academic.

Murakami, H. (2003) *Hard Boiled Wonderland and the End of the World*. London: Vintage.

Nancy, J. L. (2007). *Listening*. New York: Fordham University Press.

Nancy, J. L. (2012). 'Jean-Luc Nancy and the Listening Subject Brian Kane'. *Contemporary Music Review* 31(5–6): 439–447.

Nechvetal, J. (2011). *Immersion into Noise*. Ann Arbor, MI: Open Humanities Press.

Nietzsche, F. (1967). *The Birth of Tragedy*, trans. W. Kaufman. New York: Vintage.

North Whitehead, A. (1948). *Adventures of Ideas*. Harmondsworth: Pelican Books.

North Whitehead, A. (1957). *Process and Reality*. New York: The Free Press.

Olivier, L. (2003). 'The Past of the Present: Archaeological Memory and Time'. *Archaeological Dialogues* 10(2): 204–213.

Reynolds, S. (2005). *Rip it Up and Start Again*. London: Faber & Faber.

Reynolds, S. (2007). *Bring the Noise*. London: Faber & Faber.

Reynolds, S. (2008). *Energy Flash*. London: Faber & Faber.

Reynolds, S. (2011). *Retromania*. London: Faber & Faber.

Rojek, C. (2011). *Pop Music, Pop Culture*. Cambridge: Polity.

Ross, A. (2012). *The Rest is Noise: Listening to the Twentieth Century*. London: Fourth Estate.

Sellars, W. (1997). *Empiricism & the Philosophy of Mind*. Cambridge, MA: Harvard University Press.

Sellars, W. (2007). *In the Space of Reasons*. Cambridge, MA: Harvard University Press.

Serres, M. (1995). *Genesis*, trans. G. James and J. Nielson. Ann Arbor: University of Michigan Press.

Shaviro, S. (2012). *Without Criteria: Kant, Whitehead, Deleuze, and Aesthetics*. Cambridge, MA: MIT Press.

Shaviro, S. (2014). *The Universe of Things*. Minneapolis: University of Minnesota Press.

Sterne, J. (2003). *The Audible Past: Cultural Origins of Sound Reproduction*. Durham, NC: Duke University Press.

Sterne, J. (2011). 'The Theology of Sound: A Critique of Orality'. *Canadian Journal of Communication* 36: 207–225.

Sterne, J. (2012). *MP3: The Meaning of a Format (Sign, Storage, Transmission)*. Durham, NC: Duke University Press.

Stubbs, D. (2014). *Future Days: Krautrock and the Building of Modern Germany*. London: Faber & Faber.

Thompson, M. (2017). *Beyond Unwanted Sound: Noise Affect and Aesthetic Moralism*. New York: Bloomsbury.

Thrift, N. (2008). *Non Representational Theory: Space, Politics. Affect*. London: Routledge.

Toffler, A. (1984). *Future Shock*. New York: Bantam.

Trower, S. (2012). *Senses of Vibration: A History of the Pleasure and Pain of Sound*. London: Bloomsbury.

Virilio, P. (2005). *The Information Bomb*. London: Verso.

Voegelin, S. (2010). *Listening to Noise and Silence: Towards a Philosophy of Sound Art*. London: Bloomsbury.

Watts, M. (2001). *Heidegger: A Beginners Guide*. London: Hodder & Stoughton.

Welsch, W. (1997). *Undoing Aesthetics*. London: Sage.

Whitehead, A. N. (1929). *Process and Reality*. New York: The Free Press.

Witmore, C. L. (2006). 'Vision, Media, Noise and the Percolation of Time: Symmetrical Approaches to the Mediation of the Material World'. *Journal of Material Culture* 11: 267.

Index

CPSIA information can be obtained
at www.ICGtesting.com
Printed in the USA
LVHW091957050919
630063LV00009B/132/P

9 781501 321054